DATE DUE	
APR 18 1996	

GAYLORD   PRINTED IN U.S.A.

Chromosome Structure and Function

MICHAEL S. RISLEY, Editor
Cornell University Medical College
New York, New York

Van Nostrand Reinhold
Advanced Cell Biology Series

 VAN NOSTRAND REINHOLD COMPANY
———————————————————— New York

Copyright © 1986 by **Van Nostrand Reinhold Company Inc.**
Library of Congress Catalog Card Number: 85-17808
ISBN: 0-442-27638-9

All rights reserved. No part of this work covered by the copyrights hereon may be reproduced or used in any form or by any means—graphic, electronic, or mechanical, including photocopying, recording, taping, or information storage and retrieval systems—without permission of the publisher.

Manufactured in the United States of America.

Published by Van Nostrand Reinhold Company Inc.
115 Fifth Avenue
New York, New York 10003

Van Nostrand Reinhold Company Limited
Molly Millars Lane
Wokingham, Berkshire RG11 2PY, England

Van Nostrand Reinhold
480 Latrobe Street
Melbourne, Victoria 3000, Australia

Macmillan of Canada
Division of Gage Publishing Limited
164 Commander Boulevard
Agincourt, Ontario M1S 3C7, Canada

15 14 13 12 11 10 9 8 7 6 5 4 3 2 1

Library of Congress Cataloging-in-Publication Data
Main entry under title:
Chromosome structure and function.
   Includes bibliographies and index.
   1. Chromosomes. I. Risley, Michael S.
QH600.C495   1986   574.87'322   85-17808
ISBN 0-442-27638-9

Contents

Foreword / vii
Preface / viii
Contributors / x

1 The Structure of Interphase Chromatin, *James D. McGhee*   1

Nucleosome Core Particle, 3   The Chromatosome and Histone H1, 9   The Nucleosome, the Linker DNA, and the 10 nm "Beads-on-a-String" Filament, 13   The Chromatin Thick Fiber, 19   References, 31

2 Chromatin Structure, Gene Expression and Differentiation, *Michael Sheffery*   39

Transcriptional Regulation of Differentiation Specific Genes, 39   Structural Properties of Transcribed Genes and Active Nucleosomes, 52   Hypersensitive Sites, Active Chromatin, and Changes in Chromatin Structure During Development, Differentiation, and Gene Activation, 65   References, 79

3 Organization of Mitotic Chromosomes, *Kenneth W. Adolph*   92

Components of Metaphase Chromosomes, 92   Structural Organization of Intact Chromosomes, 103   Histone-Depleted Metaphase Chromosomes and Chromosome Scaffolds, 113   Models of Metaphase Chromosome Structure, 116   Structures of Other Chromosome Types, 119   Relationship to the Arrangement of DNA in the Interphase Nucleus, 120   References, 121

4 The Organization of Meiotic Chromosomes and Synaptonemal Complexes, *Michael S. Risley*   126

Meiosis: An Overview, 126   Synaptonemal Complexes, 130   Synaptonemal Complexes, Chromatin Loops, and Chromosome Scaffolds, 137   Synaptonemal Complexes and the Nuclear Matrix, 138   DNA Metabolism During Meiosis, 141   References, 144

5 The Lampbrush Chromosomes of Animal Oocytes, *Herbert C. MacGregor*   152

The Components of LBCs, 163   The Proteins on LBCs, 171   Expression of DNA Sequences on LBCs, 174   The Significance of Lampbrush Formation, 179   References, 182

6 Dosage Compensation in Mammals: X Chromosome Inactivation, *Leslie F. Lock and Gail R. Martin*   187

Single Active X Hypothesis, 188   Inactivation Can Be Random Or Nonrandom, 196   The Mechanism of X Inactivation, 203   Summary, 212   References, 213

7 Polytene Chromosomes, *Milan Jamrich*   221

Polytene Nuclei, 221   Molecular Basis of the Banding Pattern, 228   Bands and Genes, 230   Transcription on Polytene Chromosomes, 231   Distribution of Proteins on Polytene Chromosomes, 234   References, 236

Index /   243

Foreword

A few years ago we collected the thoughts of a number of distinguished investigators to write *Advanced Cell Biology,* which was published by Van Nostrand Reinhold. Seventy-five experts shared their insight in cell biology in what became eighty-four clear topics presented with an unprecedented timeliness. Although during the early stages it seemed too large a task to complete, the combined efforts of the talented participants resulted in that most excellent book.

The late Lazar M. Schwartz, M.D. was mainly responsible for the coordination and the success of the book, which reached a sizeable readership. Confronted with a wide variety of topics, a fast moving field, and a demanding audience, we decided next to propose to Van Nostrand Reinhold to extend the scope of the basic Advanced Cell Biology book by publishing a series of timely, concise, and didactic monographs. This Advanced Cell Biology Series will bring to investigators in the scientific community, as well as to students, an up-to-date presentation of select topics of great significance. Now we are pleased to bring our readers a new volume on this monograph series on Advanced Cell Biology.

Before Dr. Lazar M. Schwartz's untimely death, he invited Dr. Michael S. Risley to contribute to our series with a monograph on *Chromosome Structure and Function.* Now that the task has been completed, I am sure Doctor Schwartz would have been very happy with the final product.

Seven concise chapters have been written by outstanding scientists from different leading universities. They make together a point of reference that the literature will frequently refer to in days to come. Dr. Risley has certainly met all expectations regarding the quality of this volume.

MIGUEL M. AZAR, M.D., Ph.D.

Preface

The fundamental role of chromosomes in the specification and propogation of the cellular phenotype was first recognized in the late 1800s by the cytologists T. Boveri, E. Van Beneden, and W. Flemming. Their pioneering microscopic studies demonstrated that chromosome number was constant from generation to generation, that chromosomes appeared to replicate and split during mitosis, and that chromosomes were divided equally between the resultant daughter cells. Chromosome behavior during the cell cycle was also found to be highly reproducible and predictable in species such as *Ascaris*. These characteristics suggested that chromosomes may contain the genes (Mendelian factors) that Gregor Mendel had earlier postulated to be discrete packets of information for the phenotype of individuals. Linkage between genes and chromosomes was demonstrated definitively in the early 1900s by the pioneers of *Drosophila* genetics, namely T. H. Morgan, C. B. Bridges, A. H. Sturtevant, H. J. Muller, and C. Stern. These investigators coupled cytogenetic and classical genetic segregation analyses to map specific genes to individual chromosomes and chromosome regions. Biochemical studies by O. T. Avery, C. M. Macleod, and M. McCarty in 1944 subsequently demonstrated that the chromosomal macromolecule encoding genetic information was DNA since *Pneumococcus S* strain DNA could transform *R* strain *Pneumococcus* to the *S* strain. In 1953, details of the molecular organization of genetic information in DNA were provided by X-ray crystallography of DNA by M. H. F. Wilkins and co-workers, and the seminal conceptualizations of J. D. Watson and F. H. C. Crick.

Our understanding of the structure and function of chromosomes has advanced significantly during the past thirty two years, largely due to the application of electron microscopic, nucleic-acid hybridization, and recombinant DNA techniques. Major advances have been made in our knowledge of the arrangement of DNA and its associated proteins in chromatin, in the enzymology of transcription and replication, in the fine structure of genes, and in the relationships between chromatin organization and transcription. There has also been a corresponding increase in the volume of literature devoted to chromosome research, due to a large increase in the number of investigators from diverse disciplines that are now studying the biology and biochemistry of chromosomes. As a consequence, there is a need for frequent reviews of advances in chromosome research.

This book has been produced to provide a summary and critical discussion of recent research in several major areas of investigation on chromosome structure and function. Each chapter presents an introduction to the topic, a historical perspective, and a critical evaluation of recent findings. A didactic presentation is provided where appropriate. *Chromosome Structure and Function* should therefore be useful for graduate students as well as seasoned investigators.

MICHAEL S. RISLEY

Contributors

Dr. Kenneth W. Adolph
Department of Biochemistry, University of Minnesota Medical School, 435 Delaware Street, S.E., Minneapolis, Minnesota 55455

Dr. Milan A. Jamrich
Laboratory of Molecular Genetics, National Institute of Child Health and Human Development, Building 6, Room 328, Bethesda, Maryland 20205

Leslie F. Lock
Department of Anatomy, University of California at San Francisco, San Francisco, California 94143

Dr. Gail Martin
Department of Anatomy, University of California at San Francisco, San Francisco, California 94143

Dr. Herbert C. MacGregor
Department of Zoology, University of Leicester, Leicester, England, Leicester LE1 7RH, United Kingdom

Dr. James D. McGhee
Department of Medical Biochemistry, The University of Calgary Health Sciences Centre, 3330 Hospital Drive, N.W., Calgary, Alberta, Canada T2N 4N1

Dr. Michael S. Risley
Department of Cell Biology and Anatomy, Cornell University Medical College, New York, New York 10021

Dr. Michael Sheffery
DeWitt Wallace Research Laboratory, Memorial Sloan-Kettering Cancer Center, 1275 York Avenue, New York, New York 10021

1

The Structure of Interphase Chromatin

James D. McGhee
*University of Calgary,
Alberta, Canada*

The human body contains more than a billion miles of DNA. How this DNA is packaged, replicated, and transcribed is the subject of this book.

In an average cell, the meter or so of DNA is complexed with an equal mass of small basic proteins, the histones, and it is this complex that forms the structural backbone of chromatin. In addition, depending upon the cell type, there may be an equal mass of nonhistone-chromosomal proteins and a variety of different types of RNA. All this mixture is usually included in the term chromatin. This first chapter considers only the structure of the DNA-histone complex.

Although chromatin structure has been studied for decades, the major breakthrough came ten years ago with the advent of the nucleosome or subunit model for chromatin. Figure 1-1 illustrates this process of discovery by plotting the number of references to "chromatin" listed in a major indexing service, as a function of the calendar year. During the early 1970s, when the bibliographies of chromatin investigators were rapidly expanding, chromatin was being digested with nucleases, in particular with micrococcal nuclease (Clark and Felsenfeld, 1971; Rill and Van Holde, 1973; Hewish and Burgoyne, 1973; and especially Noll, 1974*a*), digested with proteases (Sahasrabuddhe and Van Holde, 1974; Weintraub and Van Lente, 1974), and examined by electron microscopy (Olins and Olins, 1973; Woodcock, 1973). Most importantly, it was becoming clear from the work of Isenberg and coworkers (reviewed in Isenberg, 1979; see also Kornberg and Thomas,

1974), that the different histone species interacted among themselves in solution to form tight stoichiometric complexes. All of these different lines of evidence were finally brought together by Kornberg (1974), who proposed that the structure of chromatin was based on a repeating subunit, a small DNA-histone complex. The term "nucleosome" was later adopted for this particle (Oudet et al., 1975). Fuller accounts of this very exciting process of discovery have been given in some of the following cited review articles.

As is evident from Figure 1-1, the annual number of references to chromatin has reached a plateau and even may be on the decline. Perhaps this indicates that the field has "matured," that is, all the easy experiments have been done. Certainly another reason for the declining publication rate (not in itself a bad thing) is the regular installments that have appeared describing the X-ray crystal structure of the nucleosome core particle at ever increasing resolution (Richmond et al., 1984; Bentley et al., 1984; Richmond et al., 1982; Klug et al., 1980; Finch et al., 1977). This fundamental kind of structural information simply renders obsolete many kinds of indirect experiments, such as chemical probing and nuclease digestion. Perhaps a third reason for the decline in interest in chromatin structure has come as backlash to the initial hopes that knowledge of the structure of the nucleosome would somehow shed light on biological processes such as the mechanism of

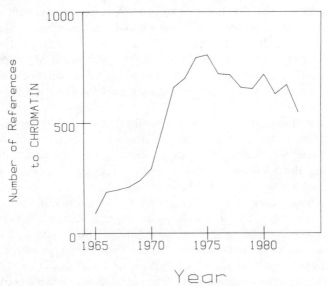

Figure 1-1. Plot of the number of references to "chromatin" as a function of time. Although the subunit model was formally proposed in 1974 by Kornberg, it is clear from the shape of the curve that many workers were anticipating a breakthrough.

differential gene activity. By and large, this has not happened, and there is as yet no clear agreement that nucleosomes have any biological function beyond DNA packing.

Figure 1-2 summarizes our present view of chromatin structure. The overall DNA-histone complex inside an interphase nucleus appears to be arranged in large loops or domains, as will be discussed in later chapters. The first reasonably highly defined level of chromatin structure, seen both by electron microscopy and by numerous biophysical techniques, is the 20-35 nm diameter chromatin thick fiber, in which the DNA has been packed and foreshortened roughly fiftyfold. This structure is stable at "physiological" salt concentrations, and particularly in the presence of divalent or multivalent cations. At very low ionic strength (on the order of millimolar), the thick chromatin fiber unwinds into the familiar "beads-on-a-string" configuration, otherwise referred to as the 10 nm filament.

The dissection of the lower levels of chromatin structure has relied heavily on the use of micrococcal nuclease, which tends to cut between the nucleoprotein particles in the so-called linker or spacer region of DNA. Such cuts release the "nucleosome," the basic repeating unit of chromatin. The nucleosome contains all the histones (one lysine rich histone, H1 or H5, and a pair each of the core histones H2a, H2b, H3, and H4) and 150-250 base pairs of DNA, depending upon the chromatin source. The linker DNA of an isolated nucleosome can be further degraded by micrococcal nuclease to give a relatively stable particle, termed the "chromatosome," which contains 165 base pairs of DNA but which still has all the histones. Further nuclease action shortens the DNA to ~145 base pairs and causes the loss of the lysine rich histone. The resulting particle, the "nucleosome core particle" is generally recognized to be the universal solution to at least the lowest level of DNA packing in all eukaryotes. As indicated, the core particle itself can be further degraded by nuclease digestion to yield small DNA or nucleoprotein fragments, whose properties reflect the internal structure of the core particle. Each of the above stages will be considered, with references and in reverse order, in the following sections. For those readers who would like more comprehensive, detailed or critical discussions of chromatin structure, the following is a partial list of reviews: Kornberg, 1977; Felsenfeld, 1978; Lilley and Pardon, 1979; Isenberg, 1979; McGhee and Felsenfeld, 1980; Mirzabekov, 1980; Igo-Kemenes et al., 1982; Cartwright et al., 1982; Hancock and Boulikas, 1982; Weisbrod, 1982; Butler, 1984; Crane-Robinson et al., 1984; Eissenberg et al., 1985.

NUCLEOSOME CORE PARTICLE

As indicated in Figure 1-2, the nucleosome core particle is the most stable nucleoprotein particle generated by micrococcal nuclease digestion of

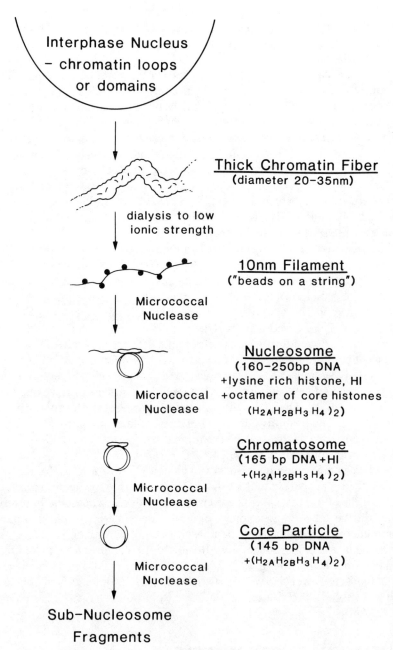

Figure 1-2. Schematic overview of chromatin structure.

chromatin. It contains 145-148 base pairs of DNA (depending on the exact extent and conditions of digestion) and two each of the histone proteins: H2a, H2b, H3, and H4. The overall molecular mass of the core particle is ~206,000 daltons, roughly equally distributed between DNA and protein. Although core particles can be produced with some difficulty, by the action of other nucleases, the preeminence of the core particle in chromatin digestion products reflects to a large degree the properties of micrococcal nuclease. It is amusing to speculate the course that chromatin research would have taken if micrococcal nuclease had never been used.

Most current and future discussion of the internal structure of the nucleosome core particle must center around the X-ray crystal structure (Richmond et al., 1984; Bentley et al., 1984; Richmond et al., 1982; Klug et al., 1980; Finch et al., 1977). The large majority of the many physical measurements, as well as the majority of the many nuclease digestion and chemical probing studies, appear to be comfortably consistent with the current X-ray structure, and by and large have been nicely incorporated and explained. This current discussion of core particle structure is thus largely a paraphrase of the recent paper by Richmond et al. (1984); the original paper should be consulted on all points of importance.

As first measured by electron microscopy and by a wide range of biophysical methods (Olins and Olins, 1973; Woodcock, 1973; Langmore and Wooley, 1975; Finch et al., 1977; Dubochet and Noll, 1978; Pardon et al., 1977; Suau et al., 1977), as well as from the low resolution crystal structure (Finch et al., 1977), the overall shape of the isolated core particle is that of a squat cylinder, ~5.5 nm high and ~11 nm across, the same relative dimensions as the average tunafish can (unpublished observations). Both neutron scattering with contrast variation (Pardon et al., 1977; Suau et al., 1977) and the accessibility of the DNA to digestion by DNase I (Noll, 1974a) argued that the DNA was wound on the outside of a protein core. Both these conclusions have been substantiated and refined by the current X-ray structure.

The 146 base pairs of DNA on the core particle surface are wound into 1.75 turns of a left-handed helix, with an average diameter of 8.6 ± 0.4 nm (Richmond et al., 1984). The superhelix pitch varies from 2.5 to 3.0 nm, depending on where on the coil the measurements are made. Thus, the two adjacent turns of DNA helix approach each other closely; the major groove of the DNA on one turn lies opposite the minor groove of the DNA on the other turn.

The duplex structure of the DNA is B-form, but is not bent completely smoothly into the core particle superhelix. Rather, there are several regions of sharper bends, at roughly 30, 60, 80, and 110 base pairs from either DNA end. The detailed stereochemistry of these bends has not yet been revealed, but apparently it does not correspond to any of the types of "kinks" that were popular several years ago. Any disruption of stacking interactions seems to be spread over several adjacent base pairs, and the phosphodiester backbone is not distorted sufficiently to be detected by ^{31}P nmr (Kallenbach et al., 1978; Klevan et al., 1979; Shindo et al., 1980). It is interesting that these regions of localized bending correspond to sites of slowest attack by DNase I (Simpson and Whitlock, 1976; Sollner-Webb and Felsenfeld, 1977; Lutter, 1978). Thus DNase I protection, so often used to probe core particle structure, does not appear to be due to straightforward steric hindrance by regions of the histones. A mild local enhancement in the rate of alkylation by dimethylsulfate (McGhee and Felsenfeld, 1979) can also be assigned to the DNA bend at 60 base pairs (Richmond et al., 1984). The resolution is as yet insufficient to determine the number of base pairs per turn (i.e., the twist or screw of the core particle DNA), but there is as yet no reason to think that the twist differs from the value of 10.0 base pairs per turn deduced from nuclease digestion studies (Klug and Lutter, 1981). This is a subject that has engaged much interest because it bears on chromatin topology.

The three-dimensional arrangement of the histones within the core particle was originally derived from electron microscopic analysis of ordered arrays of histone octamers (Klug et al., 1980). Identification of the individual histones was based heavily on the DNA-histone cross-linking experiments of Mirzabekov and coworkers (Shick et al., 1980; Belyavsky et al., 1980; Mirzabekov, 1980), histone-histone cross-linking experiments (Kornberg and Thomas, 1974; Martinson et al., 1979; Martinson and True, 1979), as well as experiments studying the roles of different histone complexes in nucleosome assembly and disassembly (Camerini-Otero et al., 1976; Bina-Stein and Simpson, 1977; Simon et al., 1978; Thomas and Oudet, 1979; Eickbush and Moundrianakis, 1978; Ruiz-Carrillo and Jorcana, 1979; Stockley and Thomas, 1979).

In the current 7 Å map, the contours of individual histones are just emerging, especially features that appear to be extended α-helical rods, but detailed histone conformations will have to await higher resolution views. Figure 1-3 (Richmond et al., 1982) shows a low-resolution view of the histone arrangement relative to the DNA. Moving along the DNA superhelix from one end of the core particle to the other, the order in which each histone is encountered is: H2a-H2b-H4-H3-H3-H4-H2b-H2a (Klug et al., 1980; Richmond et al., 1984). The (H3-H4)$_2$ tetramer assumes a central role, as had been expected from earlier studies that showed that this tetramer alone could induce nucleosome-like wrapping in DNA (Camerini-Otero et al., 1976; Bina-Stein and Simpson, 1977; Simon et al., 1978; Eickbush and Moudrianakis, 1978; Thomas and Oudet, 1979; Ruiz-Carrillo and Jorcano, 1979; Stockley and Thomas, 1979). The (H3-H4)$_2$ tetramer is shaped like a dislocated disk or horseshoe, and organizes the middle ¾ turn of nucleosome DNA. Histone

The Structure of Interphase Chromatin 7

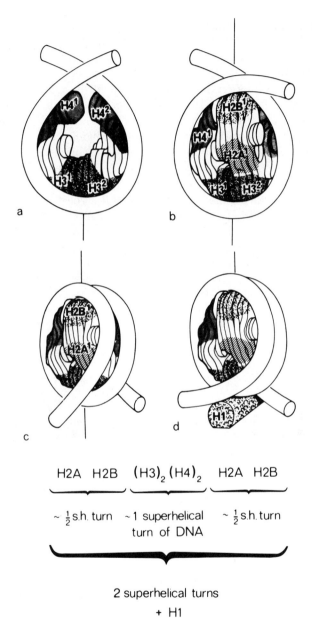

Figure 1-3. Low resolution view of the nucleosome core particle *(from Richmond et al., 1982, copyright © by Cold Spring Harbor Laboratory).* See text for full description.

H3 has now been labeled on the sulphydryl groups of Cys 110; this label, as expected (Camerini-Otero and Felsenfeld, 1977), is found to lie on the overall dyad axis of the particle (Richmond et al., 1984).

The remaining histones, H2a and H2b, have long been known to associate in solution as a dimer (reviewed in Isenberg, 1979). One H2a-H2b dimer binds to each surface of the (H3-H4)$_2$ tetramer to complete the core particle and to continue the folding of $\sim \frac{1}{2}$ turn of DNA on each side. This exterior role of the H2a-H2b dimer is consistent with its inability to fold DNA by itself (Camerini-Otero et al., 1976), with its ability to complete a core particle through rapid addition to a preformed DNA-H3-H4 complex (Camerini-Otero et al., 1976; Ruiz-Carrillo and Jorcano, 1979) and with its mode of dissociation from the isolated histone octamer (Eickbush and Moudrianakis, 1978). The composition of nuclease generated subnucleosome fragments (Nelson et al., 1982) is quite consistent with the present X-ray structure.

How is the DNA bound to the histones? Apparently the majority of histone positive charges are not involved in intimate interactions with the DNA phosphates. Perhaps as few as one or two such direct charge-charge interactions are made for every 10 base pairs of DNA, a value consistent with indirect arguments made from the ionic strength dependence of core particle melting (McGhee and Felsenfeld, 1980b). Nonetheless, it can be calculated that such a limited number of electrostatic bonds can still provide sufficient free energy to wrap the DNA (McGhee and Felsenfeld, 1980b). Neither do histone regions seem to invade the DNA grooves; this is supported by the largely unhindered reactivity of purines with dimethylsulfate (McGhee and Felsenfeld, 1979), as well as by the ability to reconstitute core particles onto fully glucosylated T4 DNA (McGhee and Felsenfeld, 1982). As Richmond et al. (1984) point out, the nucleosome core does not seem to employ a motif common to several prokaryotic sequence specific DNA binding proteins in which symmetry-paired α-helices may actually contact the edges of the DNA bases in the major groove of the specific binding site. In contrast, the path of the core particle DNA seems to be governed by a relatively small number of interactions between the phosphate backbone and a three-dimensional helical array of discrete histone binding sites. These binding sites could well be arginine residues (Ichimura et al., 1982).

The N-terminal 20-30 amino acids of each histone contain a large fraction of each histone's net positive charge. From the earliest sequence studies, these "N-terminal tails" were proposed to be the primary DNA binding sites. This has turned out probably not to be the case, since tailless histones can still fold DNA (Lilley and Tatchell, 1977; Whitlock and Stein, 1978), although these particles are, not unexpectedly, somewhat less stable. Moreover, the histone tails, even in normal core particles, seem quite mobile, especially for H2a and H2b and, at higher salt, for H3 and H4 as well (Cary et al., 1978). In with the advent of more sequence neutral cleavage reagents (Cartwright et al., 1983).

What causes nucleosomes to adopt fixed positions on a particular DNA sequence is unclear. Reconstitution of purified histones onto core-sized DNA can indeed lead to accurate positioning (Chao et al., 1979; Simpson and Stafford, 1983; Ramsey et al., 1984), suggesting that there are phasing signals in the DNA sequence. A good candidate for these signals is an anisotropic deformability, either twisting or bending, of a particular DNA sequence (see e.g., Mengeritsky and Trifonov, 1983). However, the free energies of preferential positioning are rather small, roughly 1-2 kcals favoring the most popular phase over any other phase (Ramsey et al., 1984). As Prunell and Kornberg (1982) have pointed out, these are small energy differences on which to run an important mechanism such as gene expression, and thus it is more likely that in vivo phasing arises from a boundary problem (Kornberg, 1981), that is, nucleosomes are shoved aside by the binding of a nonhistone protein at a specific DNA sequence.

We now turn to the overall conformation of the 10 nm filament, the string of nucleosomes that is the favored appearance of chromatin at very low ionic strength. Light-scattering (Campbell et al., 1978; Ausio et al., 1984), X-ray scattering (Hollandt et al., 1979), neutron scattering (Suau et al., 1979) and, of course, electron microscopy (Olins and Olins, 1973, and many examples since) all indicate that the 10 nm filament is a highly extended molecule. Indeed, the usually measured mass per unit length indicates that the linker DNA must be almost completely extended between neighboring chromatosomes. Moreover, its hydrodynamic behavior (e.g., McGhee et al., 1980, 1983*a*) is that of a flexible polymeric chain, and it must be the linker region that is imparting this flexibility. It is not clear whether any proteins, such as a part of H1 or H5 (see previous mention), are indeed bound to the linker DNA; however, even in the extended 10 nm filament configuration, cross-links can be formed between H1 molecules (Thomas and Khabaza, 1980).

The orientation of the chromatosome faces relative to the linker has been estimated by electric dichroism (McGhee et al., 1980, 1983*a*). The dichroism is strongly negative for a variety of chromatin 10 nm filaments, for example, -0.55 for chicken erythrocyte chromatin. This large negative value can only be obtained if the chromatosome faces lie close to parallel to the filament long axis, as indicated in Figure 1-4*C*. If the linker DNA is taken to be fully extended and to lie along the filament axis, the chromatosome faces can lie at most 10-20° away from parallel. This assumes that the negative limiting reduced dichroism for B-form DNA is the theoretical maximum of -1.5; if the value is as low as -1.4, due to base pair tilt or whatever (see, e.g., Lee and Charney, 1982), then the chromatosome faces must be tilted even less. These calculations were made assuming two complete smooth turns of DNA on the chromatosome surface. Would there be any effect on the calculations if the two ten base pair sections at the end of each chromatosome lay at an angle of, say, 55° from the chromatosome dyad (see previous discussion)? In the case of chicken erythrocyte chromatin (210 base pair nucleosome repeat), if the linker DNA were rigid and if it continued at the 55° exit angle all the way between neighboring chromatosomes, it can be calculated (equation 23*b* of Crothers et al., 1978) that the limiting reduced dichroism should be −0.37, significantly below the value of −0.55 actually observed. However, if only the 20 base pairs at the chromatosome termini lay at an angle of 55° to the chromatosome dyad axis and the remainder of the linker DNA thereafter lay along the filament axis, the calculated dichroism is −0.52, in complete agreement with the observed dichroism. Thus the general conclusions are unchanged: in the 10 nm filament, both the spacer DNA and the chromatosome faces must lie close to parallel to the filament axis (Fig. 1-4*C*).

Orientation of macromolecules by electric fields quite rightly raises the specter of molecular distortion. To address this question for the case of the 10 nm filament, it is important to be clear about what actually constitutes distortion. The 10 nm filament is a flexible polymer, and, in the absence of an orienting force, exists in solution as a more or less random coil, much as does DNA. In order to obtain a dichroism signal in the presence of an electric field, there must be a net internal orientation of the flexible segments relative to the orienting field; that is, in one sense the molecule has been distorted. However, we could find no evidence that, even at the highest degrees of orientation, the chromatosomes change their position relative to their linker DNAs, the real point at issue. In any case, it is comforting that flow dichroism, even in the limit of zero orienting field, also gives a negative dichroism (Baase et al., unpublished) as well as negative birefringence (Harrington, 1984). A report of a positive dichroism for low ionic strength rat liver chromatin (Tjerneld et al., 1982) remains unexplained, although the signal appears to derive from a very labile structure (Matsuoka et al., 1984).

All the biophysical measurements, in particular electric dichroism, can say nothing about the cylindrical symmetry (the disposition of chromatosomes around the filament axis). Do neighboring chromatosomes all lie on the same side of the filament, do they alternate on opposite sides, or is their relative angular orientation more complicated? Electron microscopy can potentially distinguish among these alternatives, and the general conclusion is that the 10 nm filament has a zigzag conformation, with neighboring chromatosomes lying on opposite sides of the filament axis. Thoma et al. (1979) first drew attention to this zigzag pattern of low ionic strength chromatin, and showed that the arrangement was completely dependent on the presence of histone H1. In the absence of H1, nucleosomes looked more unraveled, with the linker DNA no longer entering and exiting on the same side of the chromatosome. Worcel et al. (1981) have also described this zigzag, and have explored its topological implications.

An alternate type of zigzag configuration (Fig. 1-4D) was proposed by Worcel et al. (1981), and more recently was detected by Woodcock et al. (1984). This is perhaps more appropriately described as two-parallel-rows of stacked chromatosomes and has been most clearly detected, not in isolated chromatin at low ionic strength, but in nuclei spread on the electron microscope grid under "physiological" salt conditions (see Fig. 1-5C). Again taking Figure 1-4D at face value, the predicted dichroism would be strongly positive ($\sim ^+0.7$), clearly at variance with the dichroism measured in low ionic strength chromatin fragments. Moreover, such a molecule might be expected to be extended and rather rigid. No such component with a positive dichroism was ever detected during chromatin condensation experiments (McGhee et al., 1980, 1983a), although it is a possible explanation for the positive flow dichroism reported by Tjerneld et al. (1982; see previous mention). Both Worcel et al. (1981) and Woodcock et al. (1984) view the two-parallel-rows arrangement as a stage in the unwinding of the 30 nm chromatin fiber, and make it the basis for elegant models to be discussed in the next section. One would have to argue that the two-parallel-rows arrangement is too unstable to be detected by biophysical measurements.

THE CHROMATIN THICK FIBER

It has long been known that, under in vivo ionic conditions, chromatin does not usually exist as the 10 nm filament or string-of-beads configuration, but rather as a much thicker, somewhat irregular fiber, roughly 20-35 nm in diameter (see, e.g., Davies et al., 1974; Ris and Korenberg, 1979). Although, in this respect, the chromatin thick fiber has been studied for a longer period of time than has the structure of the nucleosome filament, it should be apparent from the discussion in the present section that the higher the level of chromatin structure, the lower the level of definitive evidence.

As mentioned previously, there have been a number of demonstrations that the 10 nm filament, at least as isolated by micrococcal nuclease digestion and dialysis to low ionic strength, can be reversibly compacted, roughly five- to tenfold, by addition of either monovalent or divalent ions. This folding transition has been monitored by sedimentation (Butler and Thomas, 1980; Thomas and Butler, 1980; Bates et al., 1981; McGhee et al., 1983b), electron microscopy (Thoma et al., 1979), light scattering (Campbell et al., 1978; Ausio et al., 1984), neutron scattering (Suau et al., 1979) and by electric dichroism relaxation times (McGhee et al., 1980, 1983a). As induced by the addition of monovalent ions, the folding transition is very broad, extending from roughly 1 mM (completely unfolded) to 100-150 mM (completely folded). Divalent ions are much more effective in inducing the compaction and, at least at very low background ionic strength, condensation is essentially complete at 1 Mg^{++} per DNA phosphate (McGhee et al., 1980, 1983a).

It is not yet clear what is the driving force behind fiber condensation. Whether it be induced by histone-histone interactions (see, e.g., Worcel and Benyajati, 1977) or induced by histone-DNA interactions (McGhee et al., 1980, 1983a; Allan et al., 1982), the added ionic strength is presumably required to shield the DNA residual negative charges, thereby allowing folding of the DNA into closer quarters. Still, even at the highest degrees of condensation, the fiber must retain a substantial negative charge.

Different methods used to investigate the thick fiber structure seem to reach remarkably different conclusions, much as in the classic tale of the blind men and the elephant. At the center of these differences are differences in the methods of chromatin preparations. The majority of biophysical studies have used chromatin fragments prepared by mild micrococcal nuclease digestion of nuclei, followed by more or less extended dialysis against a low ionic strength solvent (0.2 mM EDTA is a favorite) in order to "release" the chromatin in a soluble form. This soluble chromatin is then usually fractionated by sucrose gradient centrifugation before studying its physical behavior over a range of ionic strengths. The primary advantages of this method are that the chromatin is produced in high yield (not uncommonly >95%), histone H1 exchange (Caron and Thomas, 1981; Thomas and Rees, 1983) is inhibited, the chromatin is unaggregated, and since the transitions are reversible, the folding process can in principle be investigated rigorously. The main disadvantage of this preparative method is, of course, that the original in vivo "native" structure has been disrupted.

The alternative method of chromatin preparation is to lyse digested nuclei by prolonged dialysis into "physiological" ionic strength, followed by fractionation by sucrose gradients (again, usually maintaining elevated ionic strengths). The major motivation here is to ensure that the chromatin structure has never been disrupted, thus one hopes to be studying a structure closer to the native state. The disadvantages are that chromatin yields are usually (but not always) much lower than those obtained with the low ionic strength method; the chromatin has an increased tendency to aggregate (which would confuse many biophysical studies but not necessarily prevent meaningful electron microscopy), and metastable states are difficult to investigate. A potentially serious problem is that, at higher ionic strengths, histone H1 can and probably does exchange between different chromatin fragments (Caron and Thomas, 1981; Thomas and Rees, 1983); moreover, this H1 exchange is particularly bothersome, since it is dependent upon the length of the chromatin fragment and thus can lead to length-dependent disproportionation and hence different solubility classes of chromatin (Komaiko and Felsenfeld, 1984). The somewhat unexpected results of one biophysical study in which chromatin was prepared by fractionation on gel filtration columns at relatively high ionic strength (Fulmer and Bloomfield, 1982) could be explained in part by the loss of histone H1 that is known to occur under these conditions (Bates et al., 1981).

There is a third class of chromatin preparation methods that can be applied only to certain types of observations, but that attempts to approach even more closely to the in vivo situation. For example, nuclei have been ruptured by mild mechanical agitation, and their contents spilled onto electron microscope grids with few additional manipulations (Rattner and Hamkalo, 1978, 1979). The ultimate in gentle investigation procedures has been achieved by Langmore and coworkers, who have made a point of conducting X-ray scattering on intact nuclei, or even on intact cells (Langmore and Schutt, 1980; Paulson and Langmore, 1983). This latter technique certainly cannot be faulted for destroying the native structures, and fortunately the data support the in vivo existence of the thick chromatin fiber.

How is the DNA wound into the 30 nm thick chromatin fiber? The answer is quite unclear, but models fall into two broad classes. The first and probably more popular model for the DNA wrapping is the "solenoid" or variations thereof. In this model, first proposed by Finch and Klug (1976), the 10 nm filament is wound in a helix (whether left-handed or right-handed is unknown), with successive coils in close contact with each other. Twisted or wrapped ribbon models (Worcel et al., 1981; Woodcock et al., 1984) also can be considered as variations on a continuous helix model. Both types of models will be considered in more detail.

The second class of models proposed to describe higher order fiber structure is the "superbead" championed by a number of groups, primarily by electron microscopists (e.g., Hozier et al., 1977; Jorcano et al., 1980; Pruitt and Grainger, 1980; Subirana et al., 1981; Azorin et al., 1982; Zentgraf and Franke, 1984). This model proposes that the higher order fiber consists of a linear array of clusters of nucleosomes. The number of nucleosomes/cluster varies with the experimenter, and possibly with the chromatin source (see, e.g., Zentgraf and Franke, 1984), but generally lies in the range of 10-50 nucleosomes per superbead. There have been few detailed proposals to describe the DNA winding within a superbead (however, see Worcel, 1977), and it is not yet clear how such a bead could accommodate both variable numbers of nucleosomes and variable spacer lengths per nucleosome.

We first consider evidence in support of a superbead model for the chromatin thick fiber. Figure 1-5A, taken from Zentgraf and Franke (1984) shows one of the most convincing electron micrographs demonstrating that an impressively regular chromatin complex can indeed be produced under certain defined conditions of micrococcal nuclease digestion and sucrose gradient fractionation. In this case, the complex has been isolated from chicken erythrocyte nuclei and contains close to 20 nucleosomes/bead. Micrographs such as Figure 1-5A certainly demonstrate that regular chromatin complexes can be detected, but, as in all chromatin studies, the question remains whether this configuration exists in vivo or has been produced by the experimental manipulations.

Initially, the credibility of a superbead model suffered badly when it was shown that putative oligo-superbeads were almost certainly ribonucleoprotein particles (reviewed in McGhee and Felsenfeld, 1980a). However, this criticism is certainly not valid for later studies in which superbeads have been detected in chromatin isolated from tissues with very low RNA content (see, e.g., Zentgraf and Franke, 1984), or in which chromatin has been studied by sectioning nuclei (Subirana et al., 1981; Azorin et al., 1982). Nonetheless, oligo-superbeads have not yet been isolated in a convincing manner or yield. This is a strong and necessary prediction of such a repeating subunit model, and one would ideally like to see a sucrose gradient showing large discrete deoxynucleoprotein peaks representing multiples of superbeads. The isolation of multiple nucleosomes was one of the strongest proofs that chromatin was based on a repeating subunit model (Noll, 1974).

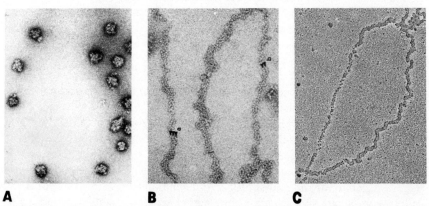

A   **B**   **C**

Figure 1-5. *(A)* Electron micrograph illustrating homogeneous "superbeads" isolated from chicken erythrocyte chromatin *(from Zentgraf and Franke, 1984, p. 272; copyright © 1984 by Rockefeller University Press; reprinted by permission)*. *(B)* A long chromatin thick fiber from cultured mouse cells, lysed into growth medium by mild mechanical agitation *(from Rattner and Hamkalo, 1978, p. 456; copyright © 1978 by Springer-Verlag; reprinted by permission)*. *(C)* View of chromatin thick fiber unraveling into a two-parallel-rows arrangement *(from Woodcock et al., 1984, p. 44; copyright © 1984 by Rockefeller University Press; reprinted by permission)*.

There are other factors that raise questions about the validity of chromatin superbeads. For example, Rattner and Hamkalo (1978) showed that relatively regular chromatin fibers seen in the electron micrographs would adopt beady and clumpy appearances simply by longer incubations, as if the beaded appearance was a metastable product of fiber unraveling (see also Pruitt and Grainger, 1980). It has not yet been demonstrated that the reverse process is a possibility, for example, a smooth fiber produced by incubation of a superbeaded fiber. Chromatin fibers analyzed in sectioned nuclei have been interpreted in terms of a superbead model (Subirana et al., 1981; Azorin et al., 1982), but the definition between superbead and interbead is rather low and, to my eye, the most striking feature of the analysis is a clear ~ 12 nm repeat along the fiber axis (see Fig. 8 of Subirana et al., 1981). Perhaps the 30 nm periodicity could equally well be interpreted in terms of a solenoid that is itself supercoiled. The tertiary coiling of the thick fiber must be borne in mind in interpreting electron micrographs, especially since the radial loops of chromatin fibers found in isolated metaphase chromosomes (see chap. 3) seem wound upon themselves (Daskal et al., 1976; Marsden and Laemmli, 1979; Adolph, 1980; Labhart et al., 1982). Such a winding of thick fibers has also been described in electron micrographs of sectioned nuclei (Andersson et al., 1984).

A slightly bothersome aspect of one particular example of superbead isolation (Jorcano et al., 1980) was that the complex did not depend on the integrity of the DNA; oligonucleosomes were also incorporated. Moreover, formation of the complex depended upon histone H1 exchange or disproportionation, and certain aspects of the complex's appearance were dependent upon the use of stain. Grau et al. (1982) have also observed that mono- and di-nucleosomes can aggregate into chromatinlike fibers.

Most biophysical measurements on isolated chromatin fragments have been interpreted as supporting a solenoid type or continuous model for chromatin, and, in some cases, as arguing against a superbead or discontinuous model. Proponents of superbeads invariably counter that their micrographs are obtained on chromatin released at ionic strengths such that the native structure has never been disrupted. However, this cannot be the whole answer. Nondigested fibers released under similar ionic conditions can show little evidence for superbeads (see Figure 1-5B). Furthermore, in both types of studies, those in which chromatin is released at low ionic strength and those in which chromatin is released at higher ionic strength, the conditions for micrococcal nuclease digestion are remarkably similar (roughly 100-150 mM monovalent ions and 1-3 mM divalent ions), and thus the overall distribution of DNA sizes should be the same in both cases. Yet, when the total DNA is purified from the digested nuclei, thus avoiding complications induced by fractionations and possible disproportionations, no evidence has been reported for a repeating fragment length in the DNA. Overall, there is the nagging suspicion that histone H1 exchange could somehow lie behind the generation of superbeads.

We now turn to a consideration of continuously wound or helical models for the thick chromatin fiber. The most popular of these models is the "solenoid" proposed by Finch and Klug (1976). The essence of this model is that the 10 nm filament is itself helically wound into a close-packed helix, thereby bringing about a further five- to tenfold compaction of the genomic DNA. This basic model has been extended by Worcel and Benyajati (1977), Worcel (1977), Thoma et al. (1979), and McGhee et al. (1980, 1983a), among others. Two detailed proposals for the DNA path are shown in Figure 1-5, Figure 1-6A, based on electron microscopy (Thoma et al., 1979), and Figure 1-6B, used to interpret our own results from electric dichroism measurements (McGhee et al., 1980, 1983a). Both models A and B are based on the premise that the 10 nm filament is described by a model such as that in Figure 1-4C above, in which the chromatosome flat faces are close to parallel to the filament axis but in which the filament remains sufficiently flexible to allow winding into a helix.

There have been two distinct variations on a continuous helix model that have been proposed (Worcel et al., 1981; Woodcock et al., 1984), both of which are based on the premise that the 10 nm fiber can adopt a zigzag ribbon configuration in which every other nucleosome packs with their faces together (see Fig. 1-4-D). This compact ribbon of two parallel rows of

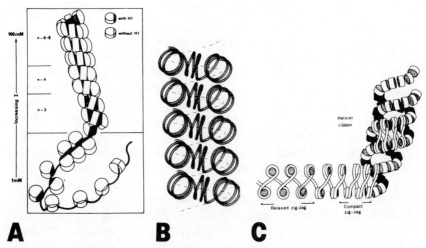

Figure 1-6. Three models for the chromatin thick fiber. (A) solenoid model after Thoma et al. (1979); (B) solenoid model after McGhee et al. (1983a); (C) helically wrapped ribbon model of Woodcock et al. (1984).

nucleosomes is then twisted (Worcel et al., 1981) or wrapped (Woodcock et al., 1984) to form the chromatin thick fiber. The model of Woodcock et al. (1984) is diagrammed in Figure 1-6C.

The original solenoid model of Finch and Klug (1976), as well as the models diagramed in Figures 1-6A and 1-6B, propose roughly six nucleosomes per turn of the secondary helix, or roughly 6 nucleosomes per 11 nm of fiber length. In contrast, model C proposes that the number of nucleosomes per turn could vary with the degree of fiber condensation, but could reach close to 12 nucleosomes per 11 nm of fiber length at the highest ionic strengths.

We first consider electron microscopic evidence in favor of a continuous helix model as opposed to a discontinuous superbead model. Figure 1-5B is a particularly impressive example of a long chromatin fiber, prepared from cultured cells broken by mild shearing under "physiological" conditions (Rattner and Hamkalo, 1978). Although the fiber clearly is not perfectly uniform, such views are more readily interpretable in terms of a continuous winding or helical model than in terms of a discrete superbead model. Thoma et al. (1979) studied isolated fragments of rat liver chromatin over a range of ionic strengths and also interpreted their pictures in terms of a simple helical winding of the 10 nm filament (Figure 1-6A), although details of nucleosome arrangement within the fiber could not be discerned. Figure 1-5C contains an electron micrograph that perhaps reveals something about the fiber internal architecture (Woodcock et al., 1984). The thick fiber, at the top and right of the figure, unravels into a thinner structure that is seen to be the ribbon of two parallel rows of stacked nucleosomes discussed previously (Fig. 1-4D). This observation, along with other evidence, forms the basis for model C (Fig. 1-6, Woodcock et al., 1984).

An extensive series of chromatin folding experiments has been conducted by Thomas and Butler (Thomas and Butler, 1980; Butler and Thomas, 1980; Bates et al., 1981). They measured the sedimentation coefficient both as a function of ionic strength and of chromatin fragment length. Chromatin fragments containing six or more nucleosomes showed an ionic strength dependent increase in sedimentation coefficient that was not found with shorter oligomers. This is the behavior expected from a solenoid type model, in which a full turn of six nucleosomes must be completed in order to stabilize the folding, and would seem more consistent with models A and B than with C (Fig. 1-6), which has no special role for six nucleosomes. The sedimentation results showed an additional curious feature that, although well documented, has not yet received an adequate interpretation. For rat liver chromatin fragments longer than ~50 nucleosomes (60 nucleosomes for chicken erythrocyte chromatin), the sedimentation coefficient undergoes an abrupt increase at 50 mM ionic strength, perhaps hinting at extra higher order folding capabilities of the thick fiber.

Light scattering (Campbell et al., 1978; Ausio et al., 1984), and dichroism relaxation times (McGhee et al., 1980, 1983a) all find that the length of the condensed chromatin fragment increases more or less linearly with increased DNA size, more consistent with a continuous helix-winding model than with a superbead model. At the highest levels of fiber condensation the measured mass per unit length corresponds to 5-7 nucleosomes per 11 nms of fiber, again favoring models A and B over C (Fig. 1-6).

Overall, the biophysical evidence, in particular the mass per unit length, tends to favor the simple solenoid geometries with six nucleosomes per 11 nm, as opposed to model C (Fig. 1-6), which has 12 nucleosomes per 11 nm at the highest degree of condensation. However, Woodcock et al. (1984) raise a very unsettling point to explain this discrepancy. As noted earlier, biophysical experiments are usually done on isolated chromatin fragments that have been refolded from low ionic strength and that could be quite heterogeneous with respect to DNA fragment length and, more importantly, with respect to imperfect fiber reconstitution. Thus, biophysical measurements are by necessity a population average, and could deviate substantially from the characteristics of a properly reassembled fiber. In the electron microscope, however, one is at liberty to select those particular conformations that appear regular or appear to have more readily interpretable structures, and then study only this subset. Thus, Woodcock et al. (1984) would argue that their mass per unit length measurements, determined on selected fibers by direct electron scattering, could be closer to the correct answer. In addition, their measurements tend to agree with model C.

The above criticism should cause the average biophysicist to have considerable anxiety. However, in defense of such physical measurements, it should be pointed out that, in many cases, sample heterogeneity does not appear to be excessive. For example, electric dichroism relaxation times of condensed chromatin fragments are acceptably first order for over 90% of the signal decay, and are no more nonlinear than would be expected from the DNA length distribution (McGhee et al., 1980, 1983a). No evidence was found for, say, 10% of the fragment population remaining unfolded. In other cases, electron microscopy of fractionated refolded fragments show quite acceptable homogeneity, as well as showing the correct axial ratio (see, e.g., Allan et al., 1981). Moreover, it is clear from Figure 1-5A that the superbeads described by Zentgraf and Franke (1984) are extremely homogeneous. Finally, Butler and Thomas showed that the sedimentation behavior of chromatin fragments that had been dialyzed to low ionic strength and then refolded was indistinguishable from the behavior of chromatin fragments that had been maintained at high ionic strength and whose structure had thus never been disrupted. In summary, the conventional mass per unit length measured for the chromatin thick fiber (roughly six nucleosomes per 11 nm) has quite rightly been questioned. However, the final decision is not yet in, and I do not think biophysical measurements should be abandoned quite yet, if only because they provide the moral satisfaction of being, in principle, objective.

We now turn briefly to a discussion of electric dichroism measurements made on condensed chromatin fragments (McGhee et al., 1980, 1983a). Although we chose to interpret our results almost exclusively in terms of a simple solenoid model (Fig. 1-6B), the measurements can obviously be used to characterize the DNA trajectory for any particular model. A brief description of the method is given in the original papers (see also Fredericq and Houssier, 1973). As noted earlier, the dichroism properties of the low ionic strength 10 nm filament show that the filament configuration is flexible and has a relatively large negative dichroism. Addition of Mg^{++} (or other polyvalent cation) causes fragment condensation (as reflected in the approximately fivefold drop in relaxation times) and a change in the way in which the DNA is wrapped (as reflected in a two- to threefold decrease in the magnitude of the negative dichroism). Over the range of 1-4 Mg^{++}/DNA phosphate, the dichroism is essentially constant and the relaxation times are constant and field independent but are dependent upon the particle size. These are the properties characteristic of rigid well-defined complexes, and no internal distortions due to the electric field could be detected. Since the condensed chromatin's dichroism remained negative (recently substantiated by the negative flow birefringence measured on similar samples; see Harrington, 1984), models are ruled out in which the flat faces of the chromatosomes are oriented perpendicular to the fiber axis. In particular, models such as Figure 1-4D, where, in the absence of further coiling, both chromatosome faces and the spacer DNA are perpendicular to the fiber axis, are completely inconsistent with our measurements.

A negative dichroism for the condensed chromatin fiber suggested that, as a first approximation, the chromatosome faces lie more or less parallel to the fiber axis. The maximum tilt possible was about 40°, and the exact angle would depend upon the disposition of the spacer DNA. More importantly, this conclusion could be combined with the cross-sectional radius of gyration measurements of Suau et al. (1979) to argue that the chromatosomes must be arranged more radially than tangentially. However, it would be difficult to estimate the uncertainty in such a statement.

To use the electric dichroism measurements to define further the DNA trajectory inside the chromatin thick fiber, an assumption had to be made about the path of the spacer DNA. We proposed that the spacer DNA was supercoiled between neighboring chromatosomes, perhaps following an extension of its superhelical path on the core particle surface. There were a number of reasons behind this proposal, but the chief motivation was to build a model that could accommodate a wide range of spacer lengths (0-80 base pairs) in roughly the same overall fiber dimensions. This assumption then allowed the simple calculation that, for chicken erythrocyte chromatin (210 base pair nucleosome repeat), the chromatosome flat faces were tilted 25° to 30° from the fiber axis. This is shown to scale in Figure 1-6B. Essentially the same measurements were performed on chromatins isolated from different sources in which spacer lengths varied from 10 to 80 base pairs. The folding and the dichroism properties of all these chromatins were quite similar and chromatosome tilt angles were estimated to lie in the range of 21° to 33°. Chromatins with longer spacers were predicted to have less chromatosome tilt, but overall, the tilt angles were remarkably independent of chromatin source and spacer length.*

Estimation of the chromatosome tilt angle associated with the simple solenoid model in Figure 1-6B could only be made, as just noted, after an assumption had been made that the spacer DNA was supercoiled. Recently, a novel technique called "photodichroism" has been introduced by Crothers and coworkers (Mitra et al., 1984), in which the chromatosome tilt angle is estimated without assumptions about spacer geometry. For calf thymus chromatin (200 base pair repeat length), Mitra et al. estimated a chromatosome tilt angle of ~33°, which is very close to the value of 30° we had previously estimated for rat liver chromatin (195 base pair repeat length). Thus the supercoiled spacer model is certainly consistent with the photochemical experiments, but of course other spacer orientations cannot be ruled out.

The same dichroism measurements that we have just been discussing can also be used to calculate the chromatosome tilt in the wrapped ribbon model (Fig. 1-6C) and the angle turns out to be roughly 30°, quite consistent with the proposed model, at least for chicken erythrocyte chromatin (see Woodcock et al., 1984). However, the dichroism results are inconsistent with the model of Worcel et al. (1981), in which the ribbon is twisted, not wrapped. It remains to be seen whether model C also is consistent with the dichroism measured with other chromatins, in particular with chromatins with either very short or very long spacer lengths. Indeed, it has not yet been demonstrated how such extreme spacer length chromatins could be accommodated into the model. In this regard, it has recently been demonstrated that brain chromatin, which has a very short nucleosome repeat length and essentially a nonexistent spacer (although spacers can be spied in the electron micrographs), can nonetheless fold into more or less normal looking chromatin fibers (Pearson et al., 1983; Allan et al., 1984). The simple

---

*Investigating solenoids by electric dichroism has unexpected drawbacks. One white-coated, bestethoscoped pathology professor informed me that he was disappointed in my seminar; he had come to learn something about chromosomes but instead it sounded more like engineering. As far as he knew, electrical solenoids were part of his car.

solenoid model in Figure 1-6B can, with some prying, still accommodate this short spacer length, but it is not clear that Figure 1-6C can, at least without some unraveling of DNA from the chromatosome surface. We have noted in the previous section that model B can easily accommodate different spacer lengths on adjacent nucleosomes, whereas model C would seem to require that adjacent spacer lengths be strongly correlated, contrary to available evidence (Strauss and Prunell, 1982).

At the opposite extreme of spacer lengths, sea urchin sperm chromatin contains a spacer DNA of roughly 80 base pairs. It has been shown how such a long spacer can easily be accommodated into model B (Fig. 1-6) by forming one full superhelical turn between neighboring chromatosomes (McGhee et al., 1983a). However, such a long spacer (~ 27 nms if extended) in the wrapped ribbon model (Fig. 1-6C) would lead to a ribbon on the order of 50 nms across, and consequently a steeper pitch angle. Such a distinctive structure should certainly be detectable by, for example, X-ray scattering or by electron microscopy. Subirana et al. (1981) have already analyzed electron micrographs of sectioned sea cucumber sperm chromatin and found no such repeating pattern along the fiber axis. However, in addition to the intensity modulations at 30 nm that were interpreted as favoring superbeads, they found a clear 11 nm repeat as expected for the simple solenoid models. In any case, it is clear that a determined study of chromatins with long spacer lengths could potentially distinguish between simple solenoid models and wrapped ribbon models.

Another potential experiment by which the different thick fiber models could be distinguished is by their topology, that is, by the supercoiling properties induced into closed circular DNA. However, it has not yet been possible to measure the topological properties of any chromatin thick fiber, and it is doubtful whether the topology of model systems such as SV40 minichromosomes is in any way related to the DNA topology of the chromatin thick fiber. Thus the model of Worcel et al. (1981), which was designed to have certain topological properties, may have been an ingenious solution to a nonproblem.

A few miscellaneous points should be made about other ways in which the models of Figure 1-6 could possibly be distinguished. Figure 1-6A, taken from Thoma et al. (1979), places the spacer DNA somewhere in the middle of the fiber, but does not define its path to the point where the model could be judged by the criteria of electric dichroism or ability to accommodate a range of spacer lengths. However, one major difference between the simple solenoid models of A and B is that A favors a histone H1 scaffold in the fiber center, whereas B predicts, as a necessary consequence of a supercoiled spacer, that H1 cannot bear any fixed relation to the fiber axis. H1 scaffolds need not be postulated to explain the existence of cross-linked H1 polymers, since these cross-linked polymers can be also generated in the extended 10 nm filament conformation (Thomas and Khabaza, 1980). Histone-DNA cross-linking experiments (Karpov et al., 1982) have been used to support a model, such as B, in which the spacer DNA is supercoiled. Woodcock et al. (1984) suggested that an occasional loss of nucleosome stacking interactions in their two parallel rows arrangement could lead to structures that have the appearance of superbeads in the electron microscope. However, it is difficult to see how such a fiber dislocation, if random, could lead to the extremely regular superbeads isolated by Zentgraf and Franke (1984). Finally, no one should have the illusion that the models shown in Figure 1-6 are the only models that have been proposed (see, e.g., Crane-Robinson et al., 1984; Nicolini et al., 1983).

What stabilizes the chromatin thick fiber? It is generally agreed that the lysine rich histones (H1 or H5, or both) are necessary to form the thick fiber. That is, polynucleosomes from which H1 has been removed do not form regular higher order structures although they can be condensed irregularly (Thoma et al., 1979). Moreover, the higher order fiber structure can be reconstituted by adding H1 back to the stripped polynucleosomes (Thoma and Koller, 1981; Allan et al., 1981). As an additional source of fiber stability, it was proposed (McGhee et al., 1980) that there could be DNA-histone interactions that could form when the fiber is condensed, but not when it is extended. One possibility, that has very attractive symmetry properties, is that the basic N-terminal histone tails (which as noted above probably play little role in maintaining the structure of the nucleosome core particle) could extend from the face of one chromatosome and bind to the DNA exposed on the surface of the neighboring chromatosome. This proposed stabilization mechanism has received some support (or, more accurately, was not ruled out) from the experiments of Allan et al. (1982), who showed that proteolytic removal of these histone tails destabilized the higher order fiber. A further attractive feature of the DNA-histone tail binding is that it suggested a mechanism whereby biologically controlled histone modifications—in particular histone acetylation, which only occurs in the histone tail region— could modulate the stability of the chromatin higher order structure. Unfortunately, in the most straightforward test of this suggestion, it was found that heavily acetylated chromatin fibers could still condense, almost indistinguishably from the unacetylated control (McGhee et al., 1983b). Thus there is currently little support for the conventional view that histone acetylation leads to chromatin decondensation, at least at the level of the thick fiber and as a result of acetylation alone. On the other hand, in certain naturally occurring examples of high levels of histone acetylation (but where there could also be other things happening) the chromatin does indeed appear to be decondensed (Christensen et al., 1984).

In summary, it is safe to say that the structure of the chromatin thick fiber remains unsolved, at least in the minds of those who have not yet proposed models. What is needed is a source of chromatin, such as unicorn epididymis, in which the thick fiber is stable, rigid, and well ordered, so that all the powerful methods of image analysis can be used. One's overall impression is that the chromatin fiber is flexible, dynamic, and even fragile, and one should not dismiss the possibility that a number of distinct conformations could peacefully coexist.

REFERENCES

Adolph, K. W. (1980). Organization of chromosomes in mitotic HeLa cells. *Exp. Cell Res.* **125,** 95-103.

Allan, J. A.; Hartman, P. G.; Crane-Robinson, C; and Aviles, F. X. (1980). The structure of histone H1 and its location in chromatin. *Nature* **288,** 675-679.

Allan, J. A.; Cowling, G. J.; Harborne, N.; Cattini, P.; Craigie R.; and Gould, H. (1981). Regulation of the higher-order structure of chromatin by histones H1 and H5. *J. Cell Biol.* **90,** 279-288.

Allan, J. A.; Harborne, N.; Rau, D. C.; and Gould, H. (1982). Participation of the core histone "tails" in the stabilization of the chromatin solenoid. *J. Cell Biol.* **93,** 285-297.

Allan, J. A.; Rau, D. C.; Harborne, N.; and Gould, H. (1984). Higher order structure in a short repeat length chromatin. *J. Cell Biol.* **98,** 1320-1327.

Andersson, K.; Bjorkroth, B.; and Daneholt, B. (1984). Packing of a specific gene into higher order structures following repression of RNA sythesis. *J. Cell Biol.* **98,** 1296-1303.

Ausio, J.; Borochov, N.; Seger, D.; and Eisenberg, H. (1984). Interaction of chromatin with NaCl and $MgCl_2$, solubility and binding studies, transition to and characterization of the higher order structure. *J. Mol. Biol.* **177,** 373-398.

Azorin, F.; Grau, L. P.; and Subirana, J. A. (1982). Supranucleosomal organization of chromatin. *Chromosoma* **85,** 251-260.

Bates, D. L.; and Thomas, J. O. (1981). Histones H1 and H5: one or two molecules per nucleosome? *Nucl. Acids Res.* **9,** 5883-5894.

Bates, D. L; Butler, P. J. G.; Pearson, E. C.; and Thomas, J. O. (1981). Stability of the higher-order structure of chicken-erythrocyte chromatin in solution. *Eur. J. Biochem.* **119,** 469-476.

Belyavsky, A. V.; Barykin, S. G.; Goguadze, E. G.; and Mirzabekov, A. D. (1980). Primary organization of nucleosomes containing all five histones and DNA 175 and 165 base-pairs long. *J. Mol. Biol.* **139,** 519-536.

Bentley, G. A.; Lewit-Bentley, A.; Finch, J. T.; Podjarny, A. D.; and Roth, M. (1984). Crystal structure of the nucleosome core particle at 16 A resolution. *J. Mol. Biol.* **176,** 55-75.

Bina-Stein, M.; and Simpson, R. T. (1977). Specific folding and contraction of DNA by histones H3 and H4. *Cell* **11,** 609-618.

Bloom, K. S.; and Carbon, J. (1982). Yeast centromere DNA is in a unique and highly ordered structure in chromosomes and small circular minichromosomes. *Cell* **29,** 305-317.

Bock, H.; Ables, S.; Zhang, X. Y.; Fritton, H.; and Igo-Kemenes, T. (1984). Positioning of nucleosomes in satellite I-containing chromatin of rat liver. *J. Mol. Biol.* **176,** 131-154.

Bonner, W. M.; and Stedman, J. D. (1979). Histone H1 is proximal to histone 2A and to A24. *Proc. Natl. Acad. Sci. (U.S.A.)* **76,** 2190-2194.

Boulikas, T.; Wiseman, J. M.; and Garrard, W. T. (1980). Points of contact between histone H1 and the histone octamer. *Proc. Natl. Acad. Sci. (U.S.A.)* **77**, 127-131.

Bryan, P. N.; Hofstetter, H.; and Birnstiel, M. L. (1981). Nucleosome arrangement on tRNA genes of Xenopus laevis. *Cell* **27**, 459-466.

Butler, P. J. G. (1984). The folding of chromatin. *C.R.C. Crit. Rev. Biochem.* **15**, 57-91.

Butler, P. J. G.; and Thomas, J. O. (1980). Changes in chromatin folding in solution. *J. Mol. Biol.* **140**, 505-529.

Camerini-Otero, R. D.; and Felsenfeld, G. (1977). Histone H3 disulfide dimers and nucleosome structure. *Proc. Natl. Acad. Sci. (U.S.A.)* **74**, 5519-5523.

Camerini-Otero, R. D.; Sollner-Webb, B.; and Felsenfeld, G. (1976). The organization of histones and DNA in chromatin: Evidence for an arginine-rich histone kernel. *Cell* **8**, 333-347.

Campbell, A. M.; Cotter, R. I.; and Pardon, J. F. (1978). Light scattering measurements supporting helical structures for chromatin in solution. *Nucl. Acids Res.* **5**, 1571-1580.

Caron, F.; and Thomas, J. O. (1981). Exchange of histone H1 between segments of chromatin. *J. Mol. Biol.* **146**, 513-537.

Cartwright, I. L.; Keene, M. A.; Howard, G. C.; Abmayr, S. M.; Fleischmann, G.; Lowenhaupt, K.; and Elgin, S. C. R. (1982). Chromatin structure and gene activity: the role of nonhistone chromosomal proteins. *C.R.C. Crit. Rev. Biochem.* **13**, 1-86.

Cartwright, I. L.; Hertzberg, R. P.; Dervan, P. B.; and Elgin, S. C. R. (1983). Cleavage of chromatin with methidium propyl-EDTA · iron(II). *Proc. Natl. Acad. Sci. (U.S.A.)* **80**, 3213-3217.

Cary, P. D.; Moss, T.; and Bradbury, E. M. (1978). High resolution proton magnetic resonance studies of chromatin core particles. *Eur. J. Biochem.* **89**, 475-482.

Chao, M. V.; Gralla, J.; and Martinson, H. G. (1979). DNA sequence directs placement of histone cores on restriction fragments during nucleosome formation. *Biochemistry* **18**, 1068-1074.

Christensen, M. E.; Rattner, J. B.; and Dixon, G. H. (1984). Hyperacetylation of histone H4 promotes chromatin decondensation prior to histone replacement by protamines during spermatogenesis in rainbow trout. *Nucl. Acids Res.* **12**, 4575-4592.

Clark, R. J.; and Felsenfeld, G. (1971). Structure of chromatin. *Nature New Biol.* **229**, 101-106.

Crane-Robinson, C.; Staynov, D.; and Baldwin, J. P. (1984). Chromatin higher order structure and histone H1. *Comments on Mol. Cell. Biophys.* (in press).

Crothers, D. M.; Dattagupta, N.; Hogan, M.; Klevan, L.; and Lee, K. S. (1978). Transient electric dichorism studies of nucleosomal particles. *Biochemistry* **17**, 4525-4528.

Daskal, Y.; Mace, M. L.; Wray, W.; and Busch, H. (1976). Use of direct current sputtering for improved visualization of chromosome topology in the scanning electron microscopy. *Exp. Cell. Res.* **100**, 204-212.

Davies, H. G.; Murray, A. B.; and Walmsley, M. E. (1974). Electron microscope observations on the organization of the nucleus in chicken erythrocytes and a superunit thread hypothesis for chromatin structure. *J. Cell. Sci.* **16**, 261-299.

Dingwall, C.; Lomonossoff, G. P.; and Laskey, R. A. (1981). High sequence specificity of micrococcal nuclease. *Nucl. Acids Res.* **9**, 2659-2673.

Dubochet, J.; and Noll, M. (1978). Nucleosome arcs and helices. *Science* **202**, 280-286.

Eickbush, T. H.; and Moudrianakis, E. N. (1978). The histone core complex: An octamer assembled by two sets of protein-protein interaction. *Biochemistry* **17**, 4955-4964.

Eissenberg, J. C.; Cartwright, I. L.; Thomas, G. H.; and Elgin, S. C. R. (1985). Selected topics in chromatin structure. *Annu. Rev. Genet.* (in press).

Felsenfeld, G. (1978). Chromatin. *Nature* **271**, 115-122.

Finch, J. T.; and Klug, A. (1976). Solenoidal model for super-structure in chromatin. *Proc. Natl. Acad. Sci. (U.S.A.)* **73**, 1897-1901.

Finch, J. T.; Lutter, L. C.; Rhodes, D.; Brown, R. S.; Rushton, B.; Levitt, M.; and Klug, A. (1977). Structure of nucleosome core particles of chromatin. *Nature* **269,** 29-36.

Fittler, F.; and Zachau, H. G. (1979). Subunit structure of the alpha-satellite DNA containing chromatin from African green monkey cells. *Nucl. Acids Res.* **7,** 1-13.

Fredericq, E.; and Houssier, C. (1973). *Electric dichroism and electric birefringence.* Clarendon Press, Oxford, England.

Fulmer, A. W.; and Bloomfield, V. A. (1982). Higher order folding of two different classes of chromatin isolated from chicken erythrocyte nuclei: a light scattering study. *Biochemistry* **21,** 985-992.

Grau, L. P.; Azorin, F.; and Subirana, J. A. (1982). Aggregation of mono- and dinucleosomes into chromatin-like fibers. *Chromosoma* **87,** 437-445.

Hancock, R.; and Boulikas, T. (1982). Functional organization in the nucleus. *Int. Rev. Cytol.* **79,** 165-14.

Harrington, R. E. (1985). Optical model studies of the salt-induced 10-30 nm fiber transition in chromatin. *Biochemistry* **24,** 2011-2021.

Hewish, D. R.; and Burgoyne, L. A. (1973). Chromatin substructure: the digestion of chromatin DNA at regularly spaced sites by a nuclear deoxyribonuclease. *Biochem. Biophys. Res. Commun.* **52,** 504-510.

Hollandt, H.; Notbohm, H.; Riedel, F.; and Harbers, E. (1979). Studies on the structure of isolated chromatin in three different solvents. *Nucl. Acids Res.* **6,** 2017-2027.

Honda, B. M.; Baillie, D. L.; and Candido, E. P. M. (1975). Properties of chromatin subunits from developing trout testis. *J. Biol. Chem.* **250,** 4643-4647.

Horz, W.; and Altenburger, W. (1981). Sequence specific cleavage of DNA by micrococcal nuclease. *Nucl. Acids Res.* **9,** 2643-2658.

Hozier, J.; Nehls, P.; and Renz, M. (1977). The chromosome fiber: evidence for an ordered superstructure of nucleosomes. *Chromosoma* **62,** 301-317.

Ichimura, S.; Mita, K.; and Zama, M. (1982). Essential role of arginine residues in the folding of DNA into nucleosome cores. *Biochemistry* **21,** 5329-5334.

Igo-Kemenes, T.; Horz, W.; and Zachau, H. G. (1982). Chromatin. *Annu. Rev. Biochem.* **51,** 89-121.

Isenberg, I. (1979). Histones. *Annu. Rev. Biochem.* **48,** 159-191.

Jorcano, J. L.; Meyer, G.; Day, L. A.; and Renz, M. (1980). Aggregation of small oligonucleotide chains into 300 Å globular particles. *Proc. Natl. Acad. Sci. U.S.A.* **77,** 6443-6447.

Kallenbach, N. R.; Appleby, D. W.; and Bradley, C. H. (1978). ^{31}P magnetic resonance of DNA in nucleosome core particles of chromatin. *Nature* **272,** 134-138.

Karpov, V. L.; Bavykin, S. G.; Preobrazhenskaya, O. V.; Belyavsky, A. V.; and Mirzabekov, A. D. (1982). Alignment of nucleosomes along DNA and organization of spacer DNA in *Drosophila* chromatin. *Nucl. Acids Res.* **10,** 4321-4337.

Klevan, L.; and Crothers, D. M. (1977). Isolation and characterization of a spacerless dinucleosome from H1 depleted chromatin. *Nucl. Acids Res.* **4,** 4077-4089.

Klevan, L.; Armitage, I. M.; and Crothers, D. M. (1979). ^{31}P NMR studies of the solution structure and dynamics of nucleosomes and DNA. *Nucl. Acids Res.* **6,** 1607-1616.

Klug, A.; and Lutter, L. C. (1981). The helical periodicity of DNA on the nucleosome. *Nucl. Acids Res.* **9,** 4267-4283.

Klug, A.; Rhodes, D.; Smith, J.; Finch, J. T.; and Thomas, J. O. (1980). A low resolution structure for the histone core of the nucleosome. *Nature* **287,** 509-516.

Komaiko, W.; and Felsenfeld, G. (1984). Solubility and structure of domains of chicken erythrocyte chromatin containing transcriptionally competent and inactive genes. *Biochemistry* **24,** 1186-1193.

Kornberg, R. D. (1974). Chromatin structure: a repeating unit of histones and DNA. *Science* **184,** 868-871.

Kornberg, R. D. (1977). Structure of chromatin. *Annu. Rev. Biochem.* **46**, 931-954.
Kornberg, R. (1981). The location of nucleosomes in chromatin: specific or statistical? *Nature* **292**, 579-580.
Kornberg, R. D.; and Thomas, J. O. (1974). Chromatin structure: oligomers of the histones. *Science* **184**, 865-868.
Kunzler, P.; and Stein, A. (1983). Histone H5 can increase the internucleosome spacing in dinucleosomes to native-like values. *Biochemistry* **22**, 1783-1789.
Labhart, P.; Koller, T.; and Wunderli, H. (1982). Involvement of higher order chromatin structures in metaphase chromosome organization. *Cell* **30**, 115-121.
Langmore, J. P.; and Schutt, C. (1980). The higher order structure of chicken erythrocyte chromosomes in vivo. *Nature* **288**, 620-622.
Langmore, J. P.; and Wooley, J. C. (1975). Chromatin architecture: investigation of a subunit of chromatin by dark field electron microscopy. *Proc. Natl. Acad. Sci. (U.S.A.)* **72**, 2691-2695.
Lee, C. H.; and Charney, E. (1982). Solution conformation of DNA. *J. Mol. Biol.* **161**, 289-303.
Leffak, I. M. (1983). Chromatin assembled in the presence of cytosine arabinoside has a short nucleosome repeat. *Nucl. Acids Res.* **11**, 5451-5466.
Lilley, D. M. J.; and Pardon, J. F. (1979). Structure and function of chromatin. *Annu. Rev. Genet.* **13**, 197-233.
Lilley, D. J. M.; and Tatchell, K. (1977). Chromatin core particle unfolding induced by tryptic cleavage of histones. *Nucl. Acids Res.* **4**, 2039-2055.
Lohr, D.; and Van Holde, K. E. (1979). Organization of spacer DNA in chromatin. *Proc. Natl. Acad. Sci. (U.S.A.)* **76**, 6326-6330.
Lohr, D.; Corden, C.; Tatchell, K.; Kovacic, R. T.; and Van Holde, K. E. (1977a). Comparative subunit structure of HeLa, yeast and chicken erythrocyte chromatin. *Proc. Natl. Acad. Sci. (U.S.A.)* **74**, 79-83.
Lohr, D.; Kovacic, R. T.; and Van Holde, K. E. (1977b). Quantitative analysis of the digestion of yeast chromatin by staphylococcal nuclease. *Biochemistry* **16**, 463-471.
Lohr, D.; Tatchell, K.; and Van Holde, K. E. (1977c). On the occurrence of nucleosome phasing in chromatin. *Cell* **12**, 829-836.
Lutter, L. C. (1978). Kinetic analysis of DNase I cleavages in the nucleosome core: evidence for a DNA superhelix. *J. Mol. Biol.* **124**, 39-420.
McGhee, J. D.; and Felsenfeld, G. (1979). Reaction of nucleosome DNA with dimethyl sulfate. *Proc. Natl. Acad. Sci. (U.S.A.)* **76**, 2133-2137.
McGhee, J. D.; and Felsenfeld, G. (1980a). Nucleosome structure., *Annu. Rev. Biochem.* **49**, 1115-1156.
McGhee, J. D.; and Felsenfeld, G. (1980b). The number of charge-charge interactions stabilizing the ends of nucleosome DNA. *Nucl. Acids Res.* **8**, 2751-2769.
McGhee, J. D.; and Felsenfeld, G. (1982). Reconstitution of nucleosome core particles containing glucosylated DNA. *J. Mol. Biol.* **158**, 685-698.
McGhee, J. D.; and Felsenfeld, G. (1983). Another potential artifact in the study of nucleosome phasing by chromatin digestion with micrococcal nuclease. *Cell* **32**, 1205-1215.
McGhee, J. D.; Rau, D. C.; Charney, E.; and Felsenfeld, G. (1980). Orientation of the nucleosome within the higher order structure of chromatin. *Cell* **22**, 87-96.
McGhee, J. D.; Nickol, J. M.; Felsenfeld, G.; and Rau, D. C. (1983a). Higher order structure of chromatin: Orientation of nucleosomes within the 30 nm chromatin solenoid is independent of species and spacer length. *Cell* **33**, 831-841.
McGhee, J. D.; Nickol, J. M.; Felsenfeld, G.; and Rau, D. C. (1983b). Histone hyperacetylation has little effect on the higher order folding of chromatin. *Nucl. Acids Res.* **11**, 4065-4075.
Marsden, M. P. F.; and Laemmli, U. K. (1979). Metaphase chromosome structure: evidence for a radial loop model. *Cell* **17**, 849-858.

Martinson, H. G.; and True, R. J. (1979). Amino acid contacts between histones are the same for plants and mammals. Binding site studies using ultraviolet light and tetranitromethane. *Biochemistry* **18,** 1947-1951.

Martinson, H. G.; True, R.; Lau, C. K.; and Mehrabian, M. (1979). Histone-histone interactions within chromatin. Preliminary location of multiple contact sites between histones 2A, 2B, and 4. *Biochemistry* **18,** 1075-1082.

Matsuoka, Y.; Nielsen, P. E.; and Norden, B. J. F. (1984). On the structure of active chromatin. *FEBS Letts.* **169,** 309-312.

Mengeritsky, G.; and Trifonov, E. N. (1983). Nucleotide sequence-directed mapping of the nucleosomes. *Nucl. Acids Res.* **11,** 3833-3851.

Mirzabekov, A. D. (1980). Nucleosome structure and its dynamic transitions. *Quart. Rev. Biophys.* **13,** 255-295.

Mitra, S.; Sen, D.; and Crothers, D. M. (1984). Orientation of nucleosomes and linker DNA in calf thymus chromatin determined by photochemical dichroism. *Nature* **308,** 247-250.

Morris, N. R. (1976). A comparison of the structure of chicken erythrocyte and chicken liver chromatin. *Cell* **9,** 627-632.

Nelson, D. A.; Mencke, A. J.; Chambers, S. A.; Oosterhof, D. K.; and Rill, R. L. (1982). Subnucleosomes and their relationships to the arrangement of histone binding sites along nucleosome DNA. *Biochemistry* **21,** 4350-4362.

Newrock, K. M.; Cohen, L. H.; Hendricks, M. B.; Donnelly, R. J.; and Weinberg, E. S. (1978). Stage specific mRNAs coding for subtypes of H2a and H2b histones in the sea urchin embryo. *Cell* **14,** 327-336.

Nicolini, C.; Cavazza, B.; Trefiletti, V.; Pioli, F.; Beltrame, F.; Brambilla, G.; Maraldi, N.; and Patrone, E. (1983). Higher-order structure of chromatin from resting cells. *J. Cell. Sci.* **62,** 103-115.

Noll, M. (1974a). Subunit structure of chromatin. *Nature* **251,** 249-251.

Noll, M. (1974b). Internal structure of the chromatin subunit. *Nucl. Acids Res.* **1,** 1573-1578.

Noll, M.; and Kornberg, R. D. (1977). Action of micrococcal nuclease on chromatin and the location of histone H1. *J. Mol. Biol.* **109,** 393-404.

Olins, D. E.; and Olins, A. Lo. (1973). Spheroid chromatin units (v bodies). *Science* **183,** 330-332.

Oudet, P., Gross-Bellard, M. and Chambon, P. (1975). Electron microscopic and biochemical evidence that chromatin structure is a repeating unit. *Cell* **4,** 281-300.

Palen, T. E.; and Cech, T. R. (1984). Chromatin structure at the replication origins and transcription-initiation regions of the riobosomal RNA genes of Tetrahymena. *Cell* **36,** 933-942.

Pardon, J. F.; Worcester, D. L.; Wooley, J. C.; Cotter, R. I.; Lilley, D. M. J.; and Richards, B. M. (1977). The structure of the chromatin core particle in solution. *Nucl. Acids Res.* **4,** 3199-3214.

Paulson, J. R.; and Langmore, J. P. (1983). Low angle X-ray diffraction studies of HeLa metaphase chromosomes: effects of histone phosphorylation and chromosome isolation procedure. *J. Cell Biol.* **96,** 1132-1137.

Pearson, E. C.; Butler, P. J. G.; and Thomas, J. O. (1983). Higher-order structure of nucleosome oligomers from short-repeat chromatin. *EMBO J.* **2,** 1367-1372.

Pruitt, S. C.; and Grainger, R. M. (1980). A repeating unit of higher order chromatin structure in chick red blood cell nuclei. *Chromosoma* **78,** 257-274.

Prunell, A. (1979). Chromatin structure: relation of nucleosomes to DNA sequence. In *Chromatin Structure and Function* (C. A. Nicolini, ed.), Plenum, New York.

Prunell, A.; and Kornberg, R. D. (1977). Relation of nucleosomes to DNA sequences. *Cold Spring Harbor Symp. Quant. Biol.* **42,** 103-108.

Prunell, A.; and Kornberg, R. D. (1982). Variable center to center distance of nucleosomes. *J. Mol. Biol.* **154,** 515-523.

Ramsey, N.; Felsenfeld, G.; Rushton, B. M.; and McGhee, J. D. (1984). A 145 base pair DNA sequence that positions itself asymetrically on the nucleosome core. *EMBO J.* **3**, 2605-2611.
Rattner, J. B.; and Hamkalo, B. A. (1978). Higher order structure in metaphase chromosome: I. The 250 A Fibre; II. The relationship between the 250 A fiber, superbeads and beads-on-a-string. *Chromosoma* **69**, 363-379.
Rattner, J. B.; and Hamkalo, B. A. (1979). Nucleosome packing in interphase chromatin. *J. Cell Biol.* **81**, 453-457.
Reeves, R.; Gorman, C. M.; and Howard, B. (1983). *In vivo* incorporation of *Drosophila* H2a histone into mammalian chromatin. *Nucl. Acids Res.* **11**, 2681-2700.
Richmond, T. J.; Finch, J. T.; and Klug, A. (1982). Studies of nucleosome structure. *Cold Spring Harbor Symp. Quant. Biol.* **47**, 493-501.
Richmond, T. J.; Finch, J. T.; Rushton, B.; Rhodes, D.; and Klug, A. (1984). The structure of the nucleosome core particle at 7A resolution. *Nature* **311**, 532-537.
Riley, D.; and Weintraub, H. (1978). Nucleosomal DNA is digested to repeats of 10 bases by Exonuclease III. *Cell* **13**, 281-293.
Rill, R.; and Van Holde, K. E. (1973). Properties of nuclease-resistant fragments of calf thymus chromatin. *J. Biol. Chem.* **248**, 1080-1085.
Ring, D.; and Cole, R. D. (1979). Chemical cross-linking of H1-histone to the nucleosomal histones. *J. Biol. Chem.* **254**, 11688-11695.
Ring, D.; and Cole, R. D. (1983). Close contacts between H1 molecules in nuclei. *J. Biol. Chem.* **258**, 15361-15364.
Ris, H.; and Korenberg, J. (1979). Chromosomal structure and levels of chromosome organization. In *Cell Biology* (D. M. Prescott and L. Goldstein, eds.), vol. 2, pp. 267-361, Academic Press, New York.
Ruiz-Carillo, A.; and Jorcano, J. L. (1979). An octamer of core histones in solution: central role of the H3-H4 tetramer in the self assembly. *Biochemistry* **18**, 760-768.
Rykowski, M.; Wallis, J.; Choe, J.; and Grunstein, M. (1981). Histone H2b subtypes are dispensable during the yeast life cycle. *Cell* **25**, 477-487.
Sahasrabuddhe, C. G.; and Van Holde, K. E. (1974). The effect of trypsin on nuclease resistant chromatin fragments. *J. Biol. Chem.* **249**, 152-156.
Shaw, B. R.; Herman, T. M.; Kovacic, R. T.; Beaudreau, G. S.; and Van Holde, K. E. (1976). Analysis of subunit organization in chicken erythrocyte chromatin. *Proc. Natl. Acad. Sci. (U.S.A.)* **73**, 505-509.
Shick, V. V.; Belyavsky, A. V.; Barykin, S. G.; and Mirzabekov, A. D. (1980). Primary organization of the nucleosome core particles. *J. Mol. Biol.* **139**, 491-517.
Shindo, H.; McGhee, J. D.; and Cohen, J. E. (1980). ^{31}P NMR studies of DNA in nucleosome core particles. *Biopolymers* **19**, 523-539.
Simon, R. H.; Camerini-Otero, R. D.; and Felsenfeld, G. (1978). An octamer of histones H3 and H4 forms a compact complex with DNA of nucleosome size. *Nucl. Acids Res.* **5**, 4805-4818.
Simpson, R. T. (1978). Structure of the chromatosome, a chromatin particle containing 160 base pairs of DNA and all the histones. *Biochemistry* **17**, 5524-5531.
Simpson, R. T. (1981) Modulation of nucleosome structure by histone subtypes in sea urchin embryos. *Proc. Natl. Acad. Sci. (U.S.A.)* **78**, 6803-6807.
Simpson, R. T.; and Stafford, D. W. (1983). Structural features of a phased nucleosome core particle. *Proc. Natl. Acad. Sci. (U.S.A.)* **80**, 51-55.
Simpson, R. T.; and Whitlock, J. P. (1976). Mapping of DNase I susceptible sites in nucleosomes labelled at 5'-ends. *Cell* **9**, 347-353.
Sollner-Webb, B.; and Felsenfeld, G. (1975). A comparison of the digestion of nuclei and chromatin by staphylococcal nuclease. *Biochem.* **14**, 2915-2920.
Sollner-Webb, B.; and Felsenfeld, G. (1977). Pancreatic DNase cleavage sites in nuclei. *Cell* **10**, 537-547.

Spadafora, C.; Oudet, P.; and Chambon, P. (1979). Rearrangement of chromatin structure induced by increasing ionic strength and temperature. *Eur. J. Biochem.* **100,** 225-235.
Stein, A; and Kunzler, P. (1983). Histone H5 can corrrectly align randomly arranged nucleosomes in a defined *in vitro* system. *Nature* **302,** 548-550.
Steinmetz, M.; Streeck, R. E.; and Zachau, H. G. (1978). Closely spaced nucleosome cores in reconstituted histone-DNA complexes and histone-H1-depleted chromatin. *Eur. J. Biochem.* **83,** 615-628.
Stockley, P. G.; and Thomas, J. O. (1979). A nucleosome-like particle containing an octamer of the arginine-rich histones H3 and H4. *FEBS Lett.* **99,** 129-135.
Strauss, F.; and Prunell, A. (1982). Nucleosome spacing in rat liver chromatin. A study with exo III. *Nucl. Acids Res.* **10,** 2275-2293.
Strauss, F.; and Prunell, A. (1983) Organization of internucleosomal DNA in rat liver chromatin. *EMBO J.* **2,** 51-56.
Suau, P.; Kneale, G. G.; Braddock, G. W.; Baldwin, J. P.; and Bradbury, E. M. (1977). A low resolution model for the chromatin core particle by neutron scattering. *Nucl. Acids Res.* **4,** 3769-3786.
Suau, P.; Bradbury, E. M.; and Baldwin, J. P. (1979). Higher-order structures of chromatin in solution. *Eur. J. Biochem.* **97,** 593-602.
Subirana, J. A.; Munoz-Guerra, S.; Martinez, A. B.; Perez-Grau, L.; Marcet, X.; and Fita, I. (1981). The subunit structure of chromatin fibers. *Chromosoma* **83,** 455-471.
Tatchell, K.; and Van Holde, K. E. (1978). Compact oligomers and nucleosome phasing. *Proc. Natl. Acad. Sci. (U.S.A.)* **75,** 3583-3587.
Thoma, F.; and Koller, T. (1981). Unravelled nucleosomes, nucleosome beads and higher order structures of chromatin: influence of non-histone components and histone H1. *J. Mol. Biol.* **149,** 709-733.
Thoma, F.; Koller, T.; and Klug, A. (1979). Involvement of histone H1 in the organization of the nucleosome and of the salt dependent superstructures of chromatin. *J. Cell Biol.* **83,** 403-427.
Thomas, J. O.; and Butler, P. J. G. (1980). Size dependence of a stable higher order structure of chromatin. *J. Mol. Biol.* **144,** 89-93.
Thomas, J. O.; and Khabaza, A. J. A. (1980). Cross-linking of histone H1 in chromatin. *Eur. J. Biochem.* **112,** 501-511.
Thomas, J. O.; and Oudet, P. (1979). Complex of the arginine rich histone tetramer with negatively supercoiled DNA. *Nucl. Acids Res.* **7,** 611-23.
Thomas, J. O.; and Rees, C. (1983). Exchange of histones H1 and H5 between chromatin fragments: a preference of H5 for higher-order structures. *Eur. J. Biochem.* **134,** 109-115.
Tjerneld, F.; Norden, B.; and Wallin, H. (1982). Chromatin structure studied by linear dichroism at different salt concentrations. *Biopolymers* **21,** 343-358.
Todd, R. D.; and Garrard, W. T. (1977). Two-dimensional electrophoretic analysis of polynucleosomes. *J. Biol. Chem.* **252,** 4729-4738.
Varshavsky, A. J.; Bakayev, V. V.; and Georgiev, G. P. (1976). Heterogeneity of chromatin subunits in vitro and location of histone H1. *Nucl. Acids Res.* **3,** 477-492.
Wallis, J. W.; Rykowski, M.; and Grunstein, M. (1983) Yeast histone H2B containing large amino terminus deletions can function in vivo. *Cell* **35,** 711-719.
Weintraub, H.; and Van Lente, F. (1974). Dissection of chromosome structure with trypsin and nucleases. *Proc. Natl. Acad. Sci. (U.S.A.)*, **71,** 4249-4253.
Weisbrod, S. (1982). Active chromatin. *Nature* **297,** 289-295.
Whitlock, J. P.; and Simpson, R. T. (1976). Removal of histone H1 exposes a 50 base pair DNA fragment between nucleosomes. *Biochemistry* **15,** 3307-3311.
Whitlock, J. P.; and Stein, A. (1978). Folding of DNA by histones which lack their amino-terminal regions. *J. Biol. Chem.* **253,** 3857-3861.
Woodcock, C. L. F. (1973). Ultrastructure of inactive chromatin. *J. Cell Biol.* **59,** 368a.

Woodcock, C. L. F.; Frado, L. L. Y.; and Rattner, J. B. (1984). The higher-order structure of chromatin: evidence for a helical ribbon arrangement. *J. Cell. Biol.* **99,** 42-52.

Worcel, A. (1977). Molecular architecture of the chromatin fiber. *Cold Spring Harbor Symp.* **42,** 313-323.

Worcel, A. and Benyajati, C. (1977). Higher order coiling of DNA in chromatin. *Cell* **12,** 83-100.

Worcel, A.; Strogatz, S.; and Riley, D. (1981). Structure of chromatin and the linking number of DNA. *Proc. Natl. Acad. Sci. (U.S.A.)* **78,** 1461-1465.

Zentgraf, H.; and Franke, W. W. (1984). Differences of supranucleosomal organization in different kinds of chromatin: Cell type-specific globular subunits containing different numbers of nucleosomes. *J. Cell Biol.* **99,** 272-286.

Zhang, X. W.; and Horz, W. (1984). Nucleosomes are positioned on mouse satellite DNA in multiple highly specific frames that are correlated with a diverged subrepeat of 9 base pairs. *J. Mol. Biol.* **176,** 105-130.

Zhang, X. Y.; Fittler, F.; and Horz, W. (1983). Eight different highly specific nucleosome phases on alpha-satellite DNA in the African green monkey. *Nucl. Acids Res.* **11,** 4287-4307.

2

Chromatin Structure, Gene Expression and Differentiation

Michael Sheffery
Memorial Sloan-Kettering Cancer Center
New York, New York

TRANSCRIPTIONAL REGULATION OF DIFFERENTIATION SPECIFIC GENES

Transcription in Isolated Nuclei

Of the many potential mechanisms regulating gene expression, induced gene transcription frequently increases the output of differentiation specific gene products (Darnell, 1982). Accurate measurement of differentiation specific gene transcription in whole cells is difficult for two reasons. First, differential nuclear and cytoplasmic RNA stabilization influences steady state RNA levels. In long labeling experiments these processes distort measurements of instantaneous synthetic rates. Short labeling times (5 min or less) are therefore required to estimate rates of nascent RNA chain growth. The necessity for short pulses leads to a second difficulty: the output from single copy, differentiation specific genes is low. Determined rates are between 10 and 2000 ppm, depending on the activity of the gene. It should be noted that ppm is a measurement of the *proportion* of RNA synthesis devoted to a particular product. During differentiation, cells often shut down peripheral RNA synthesis while differentiation specific genes are transcribed at high levels. This tends to increase the measured ppm of the differentiation specific gene even though the *rate* of transcription might remain the same. Nevertheless, for convenience and with this caveat, "rates" will be used throughout the text when referring to ppm data. Conversion of data in ppm to true synthetic rates (i.e., mass of a specific RNA produced per cell) can be performed, and this rarely changes the qualitative picture (see Landes et al., 1982). Because of low outputs, measurement of isotope incorporation after short pulses is often not practicable. For these reasons, many conclusions regarding the regulation of differentiation specific gene expression (i.e., whether regulation was exerted at the level of primary transcription or via rapid nuclear processing of constitutively transcribed sequences) remained qualified.

A suitable method for estimating the rates of nascent RNA synthesis is currently available. It relies on elongation in vitro of RNA chains initiated in vivo (Weber et al., 1977). I shall refer to this procedure as nuclear chain elongation. Nuclei are prepared and incubated in the presence of ribonucleoside triphosphates, including radioactive precursors of high specific activity. Even after short incubation times (2.5-5 min), single copy gene transcription rates are easily measured. Under conditions of the assay, RNA chain initiation is a rare event (Bitter and Roeder, 1978), so that the density of *engaged* RNA polymerase across a gene is measured. Potential chain initiations can be eliminated by performing reactions in the presence of sarkosyl, a detergent that strips unengaged polymerases and other nuclear proteins, leaving only template engaged polymerase complexes (Gariglio et al., 1981). Hybridization of purified, radiolabeled RNAs to suitable cloned DNAs (present in excess) and quantitation of RNase resistant hybrids by liquid scintillation counting accurately reflects the transcription rates of single copy, differentiation specific genes when compared to results obtained by pulse-labeling intact cells.

Global Regulation of Gene Expression by Transcription

Evidence for global regulation of gene expression by transcriptional regulation was obtained by Derman et al. (1981). cDNAs were prepared from liver mRNA, and clones carrying sequences expressed as mRNAs in other tissues (common clones) or only in liver (liver specific clones) were identified. Transcription rates of common clones were determined by pulse-labeling intact cells and by RNA chain elongation in isolated nuclei. The two methods gave comparable results when the *relative* transcription rates were compared, but nuclear chain elongation underestimated absolute rates of polymerase activity by about a factor of two when compared to pulse-labeling of intact cells. Transcription rates of liver specific clones obtained by Derman et al. (1981) were also analyzed in heterologous tissues where, by definition, little cytoplasmic RNA accumulates. Lack of mRNA expression in these tissues was found to be due largely to low transcription rates ($1/10$-$1/50$ the rate detected in liver cell nuclei). These results suggest that regulation of transcription rates has a cardinal influence on the establishment of differentiation-specific mRNA populations.

Transcriptional Regulation of Single-Copy Differentiation Specific Genes

Transcriptional Regulation and Transcriptional Domains of Murine Globin Genes. The transcription units and chromatin structure of the mouse $\alpha 1$ and Bmaj-globin genes have been described (Sheffery et al., 1984; Hofer et al., 1982). Murine globin gene transcription studies have relied largely on inducible globin gene expression in murine erythroleukemia (MEL) cells (Marks and Rifkind, 1978). After culture of MEL cells with a variety of chemical inducers, cells recapitulate many aspects of normal murine erythroid differentiation, including accumulation of α and B globins. Sensitive S1 nuclease analysis of RNAs obtained before and three days after induction shows that MEL cells exhibit more than a hundredfold increase in B-globin mRNA content, and contain greater than 10,000 copies of Bmaj-globin mRNA per cell (Wright et al., 1983). Induced cells can accumulate globin mRNA to 0.2% of total mRNA (Lowenhaupt et al., 1978).

Cloned DNA fragments spanning the Bmaj- and $\alpha 1$-globin genes have been prepared (Hofer and Darnell, 1981; Hofer et al., 1982; Sheffery et al., 1984; Fig. 2-1A), immobilized on nitrocellulose filters, and hybridized to RNA labeled in vitro by RNA chain elongation of nuclei prepared from uninduced and induced MEL cells. RNase resistance hybrids were visualized by autoradiography (Fig. 2-1B) and quantitated by liquid scintillation counting of individual dots. The results (Fig. 2-1B) show that induction of globin gene expression in MEL cells is regulated to a large degree at the level of gene transcription. Induction is ten- to twentyfold and transcription rates for the α- and Bmaj-globin genes, determined by liquid scintillation counting, are 200–400 ppm (0.04%) (Hofer and Darnell, 1981; Sheffery et al., 1984; Profous-Juchelka et al., 1983). Since cells accumulate globin mRNA to 0.2% of total mRNA, posttranscriptional mechanisms also appear to play an important role in the accumulation of globin mRNA in MEL cells (Volloch and Housman, 1981).

The transcription domains of the $\alpha 1$- and Bmaj-globin genes are also defined by the results shown in Figure 2-1B. Transcription from both genes begins predominantly near the cap site. For example, fragment αA (Fig. 2-1B) is 10 bp upstream from the $\alpha 1$ globin cap site. Little in vitro labeled RNA hybridizes to this fragment, implying few engaged polymerase molecules in this region. A similar conclusion can be drawn for transcription from fragment BA (located 370 bp upstream of the cap site; see Fig. 2-1B). Low levels of transcripts homologous to αA and BA are detected, however. For example, there is a consistent twofold increase in hybridization to fragment αA (compared to ten- to twentyfold increases to αB), suggesting that 10-20% of transcripts may have start sites in or upstream of αA. Similarly, about 20% of B transcripts appear to be homologous to fragment BA (Hofer and Darnell, 1981). Specific sites of upstream starts for the human B- and Σ-globin genes have been reported (Allan et al., 1982, 1983; Carlson and Ross, 1983), and this feature of globin gene transcriptional activation will be discussed in more detail.

It is evident from Figure 2-1B that transcription continues beyond the poly(A) addition site for both the α1- and Bmaj-globin gene. For example, fragment BF ends about 1400 bp downstream of the poly(A) addition site of

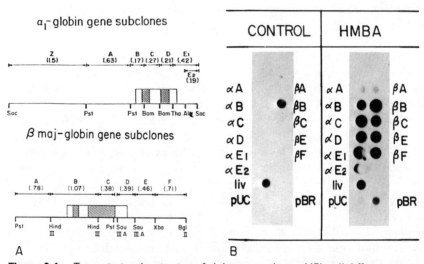

Figure 2-1. Transcriptional activation of globin genes during MEL cell differentiation. *(A)* Cloned α1- (top) and Bmaj-major (bottom) globin gene fragments. Genes are represented beginning at the cap site and ending at the poly(A) addition site. Open boxes: exons; shaded boxes: introns. Numbers in parentheses indicate approximate lengths (in kb) of cloned fragments. Restriction endonucleases used in construction of clones are indicated by letter abbreviation. *(B)* Cloned fragments, designated as in panel A, were immobilized as dots on nitrocellulose filters and hybridized to RNA prepared from 5 min pulse-labeled nuclear chain elongation reactions. An equivalent number of counts was added to each filter, and autoradiographs of washed filters are shown. Induced cells were cultured in the presence of 5 mM HMBA for 48 hr. Other DNA dots used as hybridization controls were a cDNA clone that hybridizes to nuclear RNA transcripts in both uninduced and induced MEL cells, designated "liv," and plasmids used in construction of clones *(designated pUC and pBR data from Sheffery et al., 1984, p. 422; copyright © 1984 by Academic Press [London] Ltd.).*

the Bmaj-globin gene. A region of transcription termination has been mapped 150 bp ($\pm$ 100 bp) beyond the BF fragment (Hofer et al., 1982). (A report of a precise termination site within fragment BF [Salditt-Georgieff and Darnell, 1983] was erroneous, resulting from an M13 cloning artifact that deleted a portion of fragment BF). Increased transcription is also detected in fragment $\alpha\Sigma2$, 50 bp downstream of the α1-globin poly(A) site. However, the increase is only 13% of the increase detected in the comparably sized αB fragment, suggesting that transcription may attenuate or terminate in this region (Sheffery et al., 1984).

The time course of induced transcription from the α- and Bmaj-globin genes has been investigated. A small increase in the density of engaged polymerase molecules is detected within one cell cycle (Sheffery et al., 1984; Salditt-Georgieff et al., 1984; transcription rates:20-40 ppm). Transcription rates increase slowly and continuously over the next 36-48 hrs before reaching peak rates (200-400 ppm).

Nuclear chain elongation experiments have defined several features of transcription from the murine α1- and Bmaj-globin genes. First, increases in globin mRNA in MEL cells is regulated to a large degree by increased gene transcription. Posttranscriptional control mechanisms also are implicated in contributing to the final accumulation of globin mRNA. The dimensions of the transcription unit for the Bmaj-globin gene have been well defined. Most (80%) polymerase molecules are engaged near the cap site, but 20% may be engaged further upstream. (The transcription detected further upstream is still coding strand specific; see Hofer and Darnell, 1981). The transcriptional machinery traverses 1.4 kb of Bmaj-globin exons and introns, and proceeds about 1.4 kb further downstream of the poly(A) addition site (Hofer and Darnell, 1981; Hofer et al., 1982) before terminating. Termination occurs in a defined region, probably not at a defined site. The α1-globin transcriptional domain is similar. Again, most (80%) polymerase molecules are found engaged downstream of the cap site; 20% may be engaged further upstream (again, only on the coding strand: see Sheffery et al., 1984). Transcription proceeds through the poly(A) site, but fewer polymerase molecules are engaged downstream of the poly(A) site near α1 than for a comparable fragment near the Bmaj-globin gene. It has not yet been determined if transcription terminates in this region, or proceeds further downstream. Presumably, these transcriptional domains are directly related to chromatin structure changes throughout the region. Some of these structural features will be described below.

Human Globin Gene Transcription. The detection of RNA polymerase molecules engaged upstream of the murine α- and B-globin cap sites (as in the previous discussion) suggests that transcription initiation can occur upstream of cap sites. These speculations have been confirmed for the human Σ- and B-globin genes. Allan et al. (1982) analyzed the human Σ-globin RNA produced by hemin induced K562 cells, a human leukemia cell line. They demonstrated molecules co-linear with Σ-globin mRNA initiating 215 bp upstream from the major cap site. Initiation in upstream regions accounts for 20% of all transcripts. Further analysis (Allan et al., 1983) identified at least nine initiation sites in regions as far as 4500 bp (-4500) upstream of the Σ-globin cap site. In both K562 cells and in erythroblasts purified from first trimester human embryos, 10-15% of transcripts originate in these far upstream sites. Transcripts are polyadenylated and ribosome associated. Some of these upstream sites are associated with DNase I hypersensitive sites in K562 cells (see following paragraph; Tuan and London, 1984). Interestingly, although no other nonerythroid tissues produced Σ-globin mRNA sequences, transcripts could be detected in several tissue culture cells, including HeLa cells. In these cells, transcription initiated *exclusively* at a site 215 bp upstream from the canonical cap site. Carlson and Ross (1983) also described a low level (1 part in 500) of upstream starts for the human B-globin gene in reticulocytes. These minor starts were located 235 bp upstream from the major cap site. Start sites in this region could also be detected in in vitro transcription reactions. Based on the sensitivity of the in vitro reactions to α-amanitin (synthesis was sensitive to high [100μg/ml] but not low [2 μg/ml] doses of the drug), Carlson and Ross concluded that upstream starts were pol III transcripts. Whether in vivo transcripts (which extend into globin gene coding sequences and are polyadenylated) are also pol III transcripts is not known.

Transcriptional modulation of globin genes in the human leukemia cell line K562 has also been examined (Charnay and Maniatis, 1983). K562 is induced by hemin to synthesize increased amounts of embryonic α- and B-like globin polypeptides. Low levels of adult α-globin are also produced, but no adult B-globin is detected. Analysis of globin mRNA content by Northern blotting shows that K562 cells accumulate substantial levels of globin mRNAs before induction. Culture in the presence of hemin results in a fivefold increase in globin mRNA. Nuclear chain elongation experiments also showed three- to fivefold increases in the transcription rates of the ζ-, Σ-, γ-, and α-globin genes. No transcription was detected from the adult B-globin gene. These results suggest that the increased level of globin polypeptides detected in K562 after hemin induction can be accounted for by increased gene transcription.

Developmental Regulation of Chicken Globin Genes. Changes in globin gene transcriptional activity during chicken development have also been characterized. In addition, the chromatin structure of these genes has been extensively examined (see following discussion). During chicken development, the *area opaca*, located in a posterior portion of the embryo, contains precursors of the primitive avian erythroid cells. The average generation time for cells in this area is about eight hours, and at 20-23 hrs the region is comprised of about 250,000 cells. Within 12-15 hrs (that is, at 35 hrs of development) the red cell population consists of about one million cells. Numerical considerations therefore suggest that most of the cells within the *area opaca* are within two cell cycles of terminal red cell differentiation. Clonal analysis corroborates these estimates. After explanting and dissociating cells, 85-90% of the colonies that grow (plating efficiency is 5%) stain positively for hemoglobin with benzidine. When nuclei from the precursor population are isolated and analyzed for RNA polymerase molecules transcribing the α- or B-globin genes, there is no detectable globin specific gene transcription (Groudine and Weintraub, 1981). At five days of development, nuclei isolated from red blood cells actively transcribe adult and embryonic α-globin genes and embryonic B-globin genes (Groudine et al., 1981a; Landes et al., 1982; Villeponteau et al., 1982). Globin gene expression is thus activated transcriptionally.

Between 5 and 12 days of development, a switch occurs in the globin proteins synthesized by chick red blood cells. This developmental substitution of globin polypeptides has been investigated at the transcriptional level. In five-day embryos, nuclear chain elongation labels RNA that hybridizes to two embryonic B-like globin genes termed ρ and Σ (transcription rates: 300 ppm; Landes et al., 1982; Villeponteau et al., 1982). By day 12, transcription from ρ and Σ is reduced (Groudine et al., 1981a), but not completely extinguished (Landes et al., 1982; Villeponteau et al., 1982), and adult B-globin genes (BH and B) are actively transcribed (transcription rates for B-globin reach 2000 ppm in 12-day embryos; see Landes et al., 1982).

Together, these results suggest that *determined* primitive erythroid cells do not transcribe globin genes at 23 hrs of development. By day 5, primitive erythroid cells actively transcribe embryonic, but not adult globin genes. During the switch from embryonic to adult globin protein synthesis, transcription of the embryonic genes decreases, perhaps *after* embryonic globin protein synthesis has been extinguished (i.e., there may also be posttranscriptional regulation of embryonic globin protein synthesis; see Landes et al., 1982) and adult globin gene transcription increases) Groudine et al., 1981a; Landes et al., 1982; Villeponteau et al., 1982). Thus, the activation and developmental switching of globin chain synthesis is regulated, at least in part, at the transcriptional level.

The transcriptional domains of the chicken α- and B-globin genes have also been characterized. Transcription originates predominantly near the cap site for both α- and B-globin genes (Weintraub et al., 1981; Villeponteau et al., 1982). For the adult *B* gene, transcription is also detected in a DNA fragment located more than 200 bp upstream from the predominant start site, and upstream from a nuclease hypersensitive site that maps near the gene (Villeponteau et al., 1982). Transcription is also detected downstream of the protein coding and poly(A) addition sites. Engaged polymerase molecules are detected 1500 bases downstream of one α-like gene and 600 bases downstream of the other (Weintraub et al., 1981). *B*-like globin gene transcription extends about 1 kb beyond the coding regions (Villeponteau et al., 1982).

Hormone Responsive Genes. Among the earliest single copy, differentiation specific genes analyzed by nuclear chain elongation were the hormone responsive chicken ovalbumin, conalbumin and ovomucoid genes (Swaneck et al., 1979; McKnight and Palmiter, 1979). The chromatin structure of these genes has also been extensively investigated. When stimulated by administration of estrogen, specialized cells in chicken oviduct synthesize and secrete large amounts of the major egg white proteins (ovalbumin, conalbumin, and ovomucoid). Estrogen withdrawal leads to a rapid synthetic decline. Secondary stimulation with estrogen induces a rapid increased synthesis of the major egg white proteins. The rapid secondary response, typical of steroid induced gene expression, suggests that steroid exposure results in permanent alterations in DNA templates. This is the case, and these alterations will be described in more detail. Using the nuclear chain elongation assay, McKnight and Palmiter (1979) and Swaneck et al. (1979) found that steroid induction was controlled largely by de novo transcription. For example, on secondary stimulation, the rate of ovalbumin gene transcription increased one thousandfold, from about 2 ppm to more than 2000 ppm (McKnight and Palmiter, 1979). In addition, McKnight and Palmiter (1979) also studied the time course of increased transcription during secondary response. They found a rapid increase in the rate of conalbumin gene transcription (a threefold increase within 30 minutes of estrogen administration) that paralleled appearance of estrogen receptors in the nucleus. In contrast, they observed a two-hour lag before a detectable increase in the rate of ovalbumin gene transcription, suggesting the occurrence of intermediate events between steroid receptor complex accumulation in the nucleus and gene transcription.

The transcriptional regulation of several other differentiation specific genes has been determined, and some of these are reviewed below. For the most part, the chromatin structures of these genes have not been as extensively characterized as the genes cited previously.

Liver Specific Proteins. Mouse urinary proteins (MUPs), whose primary site of synthesis is the liver, are regulated in both an age- and sex-related fashion (Derman, 1981). Little transcription of MUPs genes is detected in the livers of seven day old mice (1.4 ppm), whereas a hundredfold induction (135 ppm) of MUPs specific gene transcription is detected in adult livers. Male mice have a 5.3-fold higher steady state level of MUPs mRNA than female mice. Nuclear chain elongation experiments show that this difference can be accounted for by differences in the transcription rates of MUPs genes in male versus female livers (adult males:127 ppm; female:18ppm).

The mouse liver proteins albumin and α fetoprotein have also been analyzed for their transcriptional regulation during development (Tilghman and Belayew, 1982). Albumin and α fetoprotein mRNAs accumulate in embryonic mouse livers. Near birth, α fetoprotein mRNA falls rapidly and is undetectable 14 days postpartum. Albumin mRNA continues to accumulate. At least part of the reduction in α fetoprotein mRNA concentration is regulated at the transcriptional level, and parallels the decrease in liver cell mitotic index. Interestingly, α fetoprotein synthesis also increases during liver regeneration or after other insults that cause liver cells to enter the cell cycle. These latter results suggest DNA replication may be involved in increased α fetoprotein gene transcription. (Replication has also been implicated in the periodic expression of the yeast histone genes; see Hereford et al., 1982.)

Parotid Gland Specific Proteins. The mouse α amylase gene is transcribed in both liver and parotid glands. Two distinct mRNAs are produced in the two tissues. They have identical coding but different upstream noncoding sequences and cap sites (Young et al., 1981). Parotid gland transcripts are initiated at a promoter over 2 kb upstream of the liver promoter, and parotid gland transcripts accumulate to at least one hundredfold higher concentration than their liver counterparts. Schibler et al. (1983) showed that the α amylase gene in the parotid gland was transcribed about 30 times more efficiently than the gene in the liver. The increased rate of transcription is probably accounted for by a strong promoter accessible only in parotid cells.

RNase treatment of nuclei before nuclear chain elongation allowed a direct estimate of the rate of chain growth in vitro. About 60 base pairs of preexisting RNA chains are protected, probably by RNA polymerase. At various times during nuclear chain elongation, purified RNA was analyzed for length distribution on agarose gels followed by autoradiography. Chain elongation rates were estimated to be about 10 nucleotides per min in nuclei obtained from both parotid gland and liver cells.

Quantitation of hybridization to restriction fragments containing the α amylase gene showed that transcription of the gene in both parotid and liver nuclei initiates predominantly near the cap site of each gene (Schibler et al., 1983). Interestingly, the gene is not transcriptionally silent in all heterologous tissues. Nuclear chain elongation experiments show α amylase transcription (at low levels) in brain nuclei (brain does not accumulate α amylase mRNA). These transcripts, unlike transcripts found in the parotid gland and the liver, appear to be part of transcripts initiated several kb further upstream. This result raises several interesting issues. These transcripts may be analogous to aberrantly initiated E-globin gene transcripts found in nonerythroid cells (Allan et al., 1983). Alternatively, they might suggest that inefficient polymerase II termination can lead to transcription of genes normally not expressed as mRNAs. Lastly, their transcription suggests that, in the brain, α amylase is contained in transcriptionally "competent" chromatin. Since the α amylase promoter is probably not used, however, tissue specific regulatory factors are probably inactive or lacking.

Critical Regions Required for Differentiation Specific Gene Expression

How can the critical DNA sequences required for the activation of differentiation specific genes be determined? One approach to this problem has been to introduce genes into heterologous or homologous host cells via DNA transfection (Wigler et al., 1979).

The activity of the α- and B-globin genes has been well characterized in transient expression assays after transfection into heterologous cell systems, and by analysis of stable transfectants in homologous (erythroleukemia) cell lines. These observations represent a continuous line of investigation, and will be taken as a model for the in vitro genetic dissection of sequences required for differentiation specific gene expression.

Sequence analysis of globin genes reveals three conserved sequences upstream from the cap sites: a "distal" region from -83 to -111, a region from -67 to -77 (referred to for convenience in the rest of the text as the CCAAT box) and a region at -24 to -32 (referred to as the TATA box; cap site $= +1$; Efstradiadis et al., 1980; Dierks et al., 1983). In vitro transcription of B-like globin genes has shown that an intact TATA box is the minimum upstream sequences required for correct transcript initiation in vitro. (Grosveld et al., 1981; interestingly, sequences between -700 to -1000 can have an inhibitory effect on in vitro transcription reactions; see Gregory and Butterworth, 1983). The minimum upstream sequences required for α-globin gene transcription in vitro are also reported to include only the TATA box (Talkington and Leder, 1982).

Several differences, however, in the behavior of the α- versus B-globin promoters from various species have been determined in transient expression assays using both replicating and nonreplicating vectors. When assayed in replicating systems (i.e., vectors containing an SV40 origin of replication transfected into COS cells), the α promoter alone actively directs correct initiation of α-globin-containing transcripts (Humphries et al., 1982; Mellon et al., 1981). Transfected cells accumulate up to 200,000 copies of α-globin mRNA per cell within 48 hrs. Template copy number is estimated to be 200,000 to 400,000 copies per cell so that, on average, about one properly initiated α-globin transcript is produced per template in 48 hrs (Mellon et al., 1981). Deletions show that the minimum sequence requirements for α-globin transcription in transient assays includes the CCAAT box as well as the TATA region for correct initiation. In marked contrast, although abundant transcripts containing human B-globin gene sequences are produced in replicating systems, virtually none of the transcripts initiates at the cap site. Correct initiation is detected, however, if the vector includes the heterologous SV40 enhancer element. When linked to an enhancer element (Humphries et al., 1982; Banerji et al., 1981; Grosveld et al., 1982; Dierks et al., 1983), a region including the CCAAT box is required for efficient B-globin transcript initiation.

In nonreplicating transient expression systems correct transcription from the human B-globin gene also shows a dramatic (over one hundredfold) stimulation by linkage to an enhancer sequence; in nonreplicating vectors, expression of the human α-globin gene is also boosted five- to tenfold by linkage to an enhancer (Treisman et al., 1983a). Comparison of the level of correctly initiated transcripts from the human α-globin gene in replicating versus nonreplicating systems shows fiftyfold more α-globin transcripts after replication, an increase in expression readily accounted for by increased template number. Interestingly, comparison of B-globin transcripts produced by enhancer containing vectors in replicating versus nonreplicating systems shows that the amount of correctly initiated B-globin transcripts is the same even though templates are highly amplified by replication (Treisman et al., 1983a; incorrectly initiated B-globin transcripts are greatly amplified in the replicated vectors). This surprising result suggests that cellular factors required for enhancer dependent B, but not for the enhancer independent α-globin gene transcription may be limiting, or that vector replication itself interferes with enhancer dependent transcription (Treisman et al., 1983a).

Unexpectedly, the dependence of B-transcription on a *cis* linked enhancer is relieved in *trans* by the product of the adenovirus E1A region (E1A also eliminates any effect of an enhancer element on stimulating correct initiation from the α promoter). A naturally occurring C-G transversion at -87 in the B-globin gene (which leads to a tenfold decrease in enhancer dependent B-globin transcription; Treisman et al., 1983b) has no effect on E1A mediated transcription. However, an A-G transition at -28, within the TATA box, results in a tenfold decrease in E1A mediated transcription, regardless of the presence of an enhancer (Green et al., 1983; Treisman et al., 1983a). These results suggests that E1A mediates its effects through the TATA box, and that alternative pathways for transcription complex formation might exist.

Together, these results show several differences between two differentiation specific promoters that are coordinately regulated in vivo. Transient assays in heterologous cells show that B-like globin genes are dependent on a *cis* acting enhancer. It is possible that, in vivo, B-globin gene expression also depends on an (erythroid specific?) enhancer. The α promoter is not as dependent on enhancers in transient assays. It can be speculated that either initiation from the α-globin does not require transcription factors that might be locally concentrated by enhancers, or that a similar function is performed in heterologous cells by sequences located near or within the α-globin gene.

It is too early to coherently describe how initiation complexes are assembled. A cascade model, dependent on a series of protein-protein interactions (Travers, 1983) might be one way to concentrate and assemble factors required for initiation. Particularly strong DNA-protein binding interactions (that might also be sensitive to chromatin structure) or high concentrations of specific initiation factors might enable certain "steps" in assembly to be circumvented. One possibility is that the B-globin gene expression requires a transcription factor that a heterologous enhancer element traps and concentrates locally (Scholer and Gruss, 1984). Based on deletion studies, this factor would mediate its action through sequences near the CCAAT box. Its role might be to bind factors that also interact with the TATA box. The α-globin gene might either have an enhancer-like activity or a CCAAT box that can interact directly with initiation factors. Alternatively, α-globin might be able to dispense with this factor in initiation complex formation. E1A-like proteins appear to interact directly with the TATA box region, an interaction that might obviate requirements for factors enriched by enhancers. The E1A protein might be able to interact directly with the TATA box to stimulate initiation, either because of its high concentration or because of unusual binding properties.

Transient assays conducted in heterologous cells make it impossible to study sequences involved in tissue specific regulation. A second class of studies has, therefore, focused on analyzing globin genes stably introduced into erythroid cell lines. A suitable host for studying adult globin gene regulation has been the MEL cell, since such cells transcriptionally regulate endogenous adult globin genes. Chimeric genes comprised of mouse B-globin upstream regulatory and coding sequences linked to human B-globin downstream coding and flanking sequences, or intact human globin genes (with or without SV40 enhancer sequences) have all been transfected and stably integrated into MEL cells (Chao et al., 1983; Wright et al., 1983; Spandidos and Paul, 1982). The exogenously introduced globin genes are often inducible in MEL cell transformants. Indeed, B-globin genes that require enhancers for activity in transient assays are readily inducible when introduced into MEL cells (Chao et al., 1983; Wright et al., 1983).

Regulation of adult but not fetal or embryonic B-like globin genes is observed in MEL cells (Wright et al., 1983; again, MEL cells regulate their adult genes). At most, 1.5 kb of upstream human flanking sequences and 1.2 kb of mouse upstream flanking sequences are required for regulation (Wright et al., 1983; Chao et al., 1983). Induced expression of the mouse-human chimeras was shown to be at the level of increased gene transcription by nuclear chain elongation experiments. However, quantitative differences are observed in the expression of the transfected genes. Nuclear transcription experiments show significantly less transcription from transfected chimeric mouse/human globin genes than from the endogenous mouse globin genes (Chao et al., 1983), and transfected cells accumulate ten- to one hundredfold less mRNA than the endogenous mouse genes. Transfected genes are thus regulated, but are weakly expressed when compared to endogenous genes. In general, transfected genes appear to be readily expressible and in an "open" chromatin configuration that may render them accessible to cellular transcription factors (Weintraub, 1983). However, the quantitative differences observed in expression of transfected genes suggests that there might be some feature of chromatin structure near the transfected genes that affects their expression. These differences have not yet been reported.

More recent transfection results underline differences between α- and B-globin gene regulation. As described above, when human B-globin genes are transfected into MEL cells they are regulated like the endogenous murine globin genes. That is, they are not expressed in uninduced cells, but are expressed after induction. In contrast, human α-globin genes transfected into MEL cells are not regulated; that is, they are expressed constitutively (Charnay et al., 1984). If transfected genes are generally assembled into an "active" configuration, as discussed above, these results imply that the α-globin gene can utilize generalized transcription factors in the active chromatin configuration, whereas a second induction dependent event is required to activate transfected B-globin genes. (The chromatin structure of the endogenous α-globin gene changes during MEL cell induction; see Sheffery et al., 1984. Presumably the structure of the endogenous α-globin gene before induction prevents interaction with generalized transcription factors.)

Interestingly, analysis of hybrid genes suggests that some regulatory sequences are located downstream of the cap site of both globin genes. Hybrids comprised of α-globin promoter fused to the body of the B-globin gene are regulated in MEL cells after transfection (i.e., they are off before and on after induction, behaving like transfected B-globin rather than α-globin genes). In contrast, fusions of the B promoter to the a body are constitutively expressed (Charnay et al., 1984). In addition, murine H-2 genes are constitutively expressed after transfection into MEL cells. However, fusions comprised of the H-2 promoter attached to the body of the B-globin gene, or of the B promoter linked to the H-2 body are both regulated during MEL cell induction (Wright et al., 1984). Together, these results imply that regulatory elements for globin gene expression are located both upstream and downstream of the mRNA cap site.

In summary, transfection studies suggest dramatic differences in the regulation of promoters controlling two closely related, coordinately expressed, differentiation specific genes. They suggest that some sequences required for tissue specific regulation are located within 1.2-1.5 kb of the cap sites. These sequences, regardless of their species origin, can interact with factors required for the regulation of globin gene expression during MEL cell differentiation. This regulation is exerted, at least in part, at the level of increased gene transcription. Other sequences, located downstream of the mRNA cap site, also play important roles in regulating the expression of the globin genes. Indeed, the effects of the internal control elements can dominate elements upstream of the cap site. Expression of stably transfected genes, even in homologous cell types, is low. Perhaps larger DNA regions must be introduced to achieve structures capable of assembly into local chromatin environments that can support maximal gene expression. Alternatively, it may be difficult for DNA inserted into heterologous chromosomal positions to spontaneously co-opt all of the structures required for maximal gene expression, even though transfected genes have upstream hypersensitive sites and, at least by this criterion, are in an active configuration.

STRUCTURAL PROPERTIES OF TRANSCRIBED GENES AND ACTIVE NUCLEOSOMES

Before transcription complexes can assemble to begin transcription from differentiation specific genes, a local restructuring of chromatin must have occurred to provide a transcriptionally "competent" template. What distinguishes transcriptionally competent templates from bulk DNA? Active (some prefer "dynamic": see Ryoji and Worcel, 1984) chromatin is distinguished from bulk (or static) chromatin by several biochemical criteria. An altered structure has been inferred from a general sensitivity to DNase I, a feature that has stimulated direct searches for proteins that confer sensitivity in vitro. Small regions near active genes (hypersensitive sites) are exquisitely sensitive to nucleases and other reagents. This observation has led to investigations of local DNA sequence features detectable in vitro that might mediate hypersensitivity in vivo. A battery of methyl-sensitive restriction endonucleases has also permitted limited comparisons of methylation patterns near active and inactive genes. These biochemical approaches to characterizing competent templates have produced an enormous amount of information. Only a fraction of the literature can be reviewed here. Curiously, several fundamental issues remain somewhat obscure, and the literature is replete with apparently contradictory results. Some of these apparent difficulties will be discussed. Out of self-interest, examples will be drawn from investigations of the chromatin structure of murine globin genes. The reader should thus be made aware of recent reviews where different examples and points of view can be found (Igo-Kemenes et al., 1982; Weisbrod, 1982a).

Accessibility of Active Chromatin to DNase I

Some structural feature makes active genes readily accessible to digestion by DNase I. This dramatic difference in accessibility, originally observed by Weintraub and Groudine (1976), has been demonstrated for many active or *potentially* active genes. These include globins (Weintraub and Groudine, 1976; Stalder et al., 1980a, 1980b; Miller et al., 1978; Young et al., 1978; Sheffery et al., 1982; Groudine et al., 1983), ovalbumin (Bellard et al., 1980; Lawson et al., 1980, 1982; Shepherd et al., 1980), immunoglobulin (Storb et al., 1979, 1981; Mather and Perry, 1983), vitellogenin (Burch and Weintraub, 1983), insulin (Wu and Gilbert, 1981), histone (Samal et al., 1981), protamine (Levy-Wilson et al., 1980), and ribosomal (Stalder et al., 1978) genes.

In these experiments nuclei are isolated and incubated in the presence of DNase I. Purified DNAs from the digests are assayed for the concentration of specific genes. Usually this is done semiquantitatively by cleaving purified DNAs to completion with a restriction endonuclease, sizing the digests by electrophoresis, transferring to nitrocellulose, and probing the immobilized DNA with cloned DNAs radiolabeled in vitro. An example of this procedure, illustrated in Figure 2-2, shows that the *B*maj-globin gene in uninduced MEL cells (conditions where the gene is not actively transcribed) is more sensitive to digestion by DNase I than a gene not expressed in the erythroid lineage (an immunoglobulin gene; see also Miller et al., 1978). The enhanced susceptibility of the *B*maj-globin gene in MEL cells is thus not dependent on its transcription per se. It appears to reflect the commitment of this erythroid cell line to globin gene expression. DNase I sensitivity can thus precede gene transcription (see also Stalder et al., 1980a, and Storb et al., 1981). Sensitivity can also persist after transcription has ceased (Weintraub and Groudine, 1976; Young et al., 1978).

Interestingly, the DNase I sensitivity of the *B*maj-globin gene in MEL cell increases after transcriptional activation of the gene (Hofer et al., 1982; Smith and Yu, 1984). Enhanced sensitivity of the transcribed domains has also been described for the chicken *B*-globin cluster (Stalder et al., 1980*b*). In this latter case, it was shown that actively expressed coding regions are highly DNase sensitive, and are flanked by sequences more sensitive to digestion than a neutral, nonexpressed gene, but less sensitive than coding regions. The intermediate level of sensitivity in flanking regions extends at least 8 kb downstream and 6–7 kb upstream from the *B*-globin gene cluster, where it ends abruptly. The gene cluster itself spans 11 kb, so the sensitive domain is at least 25 kb.

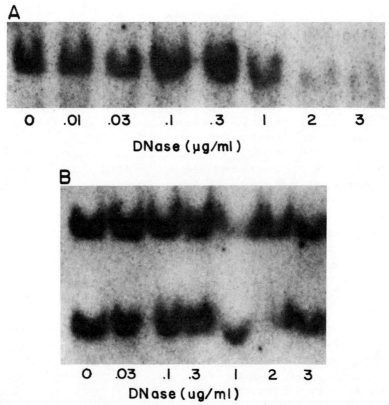

Figure 2-2. Relative sensitivity of chromatin containing the *B*maj-globin gene and an immunoglobulin gene to digestion by DNase I in uninduced MEL cells. Nuclei from uninduced cells were purified and digested with the amounts of DNase I indicated below each lane. Purified DNA was cleaved with Eco RI and analyzed for the presence of the *B*maj-globin gene *(A)*, or the Igα gene *(B)*. *(From Sheffery et al., 1982)*

The dimensions of the sensitive domain near the ovalbumin locus in chicken oviduct nuclei have also been reported (Lawson et al., 1980, 1982). Ovalbumin and adjacent steroid inducible genes are in a sensitive domain spanning about 100 kb; sensitivity falls off gradually toward the ends. The sizes of DNase I sensitive domains (at least 25 kb and as large as 100 kb) are consistent with and suggest a relationship to lampbrush chromosome loops (Gall and Callan, 1962) or to loops observed after deproteinization of metaphase chromosomes (Paulson and Laemmli, 1977; see chap. 3). Within these large sensitive domains, coding regions show an even greater sensitivity to DNase I. This level of sensitivity might be related to their transcription rate or to associations with specific proteins (HMGs 14 and 17).

The increased DNase I sensitivity of a large domain is consistent with models in which a primary event in gene activation is the selective unfolding of a chromosomal loop containing differentiation specific genes. Increased DNase I sensitivity might be a consequence of the unfolding. Evidence that active domains are contained in unfolded conformations has been presented (Kimura et al., 1983). In the open conformation, transcription factors might then be able to interact with specific regulatory sequences or structures. These interactions might further alter compaction of the coding regions in preparation for transcription.

Is domain unfolding related to cellular determination? One set of experiments suggests that it is not. Two cell cycles before terminal differentiation, the globin genes in a nearly pure population of *determined* chick erythroid precursor cells are not sensitive to DNase I (they are also not transcribed, are highly methylated, and exhibit no hypersensitive sites; see Groudine and Weintraub, 1981). These results suggest that establishment of DNase I sensitivity, while it must be an early event in gene activation, need not be related to determination.

Accessibility of Active Sequences to Micrococcal Nuclease

Micrococcal nuclease cuts bulk chromatin between nucleosomal linkers, releasing particles containing histones and DNA. When DNA purified from digests is analyzed by electrophoresis, a characteristic ladder of bands is generated. The bands have lengths that are integral multiples of the DNA associated with a nucleosome monomer (about 200 bp; see Sollner-Webb and Felsenfeld, 1975; Axel, 1975). The digestion pattern is diagnostic of nucleosomal organization. Are actively transcribed genes (more sensitive to digestion by DNase I than bulk DNA by a factor of 10; see Wood and Felsenfeld, 1982) more sensitive than bulk DNA to digestion by micrococcal nuclease? Several reports show that genes readily accessible to DNase I are not preferentially degraded by micrococcal nuclease. This has been demonstrated for the globin genes in chick erythrocytes (Weintraub and Groudine, 1976), and for the actively transcribed ovalbumin genes in hen oviduct (Bloom and Anderson, 1978; Bellard et al., 1977; Garel and Axel, 1976). By contrast, other reports suggest that the ovalbumin genes in oviduct are more sensitive to micrococcal nuclease digestion than bulk DNA (Senear and Palmiter, 1981), and that chick globin genes in erythrocytes are about three times more sensitive than bulk DNA (Wood and Felsenfeld, 1982). Actively transcribed ribosomal RNA genes (Reeves and Jones, 1976; Reeves, 1978; Mathis and Gorovsky, 1977; Johnson et al., 1978) and the Bmaj-globin genes in MEL cells (Smith and Yu, 1984) are also reported to be more sensitive to micrococcal nuclease digestion than are inactive sequences. In the case of the mouse globin genes, sensitivity was directly related to transcription rate (Smith and Yu, 1984). At least some of these discrepancies can be accounted for by methodological differences (see Senear and Palmiter, 1981). Together the data suggest that transcription units are more susceptible to degradation by micrococcal nuclease, and that susceptibility might depend on transcription rate.

Micrococcal nuclease can also be used to readily differentiate another feature of actively transcribed genes from inactive sequences. In the presence of divalent metal ions, micrococcal nuclease preferentially solubilizes monomers from active transcription units (chicken ovalbumin: Bloom and Anderson, 1978, 1979; Bellard et al., 1977; Senear and Palmiter, 1981; chick globin: Bloom and Anderson, 1979; Nicolas et al., 1983; expressed trout testis sequences: Levy et al., 1979). After brief digestions (2% acid solubility), monomers solubilized in 5mM Mg (5% of total DNA) can be 6.5-fold enriched in active sequences (i.e., the solubilized monomers contain almost 35% of all active sequences; Bloom and Anderson, 1979). Preferential solubilization of monomers can extend into upstream noncoding regions (Wood and Felsenfeld, 1982; Nicolas et al., 1983), and can be even greater than for coding regions (Nicolas et al., 1983). Gottesfeld and Butler (1977) have also shown that active nucleosomes can be selectively enriched by precipitation of inactive monomers with 2mM Mg after digestion with DNase II, an enzyme that preferentially cleaves between nucleosomes.

Examination of proteins associated with the solubilized monomers shows that they are enriched in HMGs 14 and 17 or closely related nonhistone proteins (Levy et al., 1979; Goodwin et al., 1979, 1981). This suggests that active nucleosomes may also be enriched in HMGs 14 and 17. It is also possible that these proteins help solubilize monomers under these conditions. Indeed, extended micrococcal nuclease digests in the presence of Mg containing buffers produces a biphasic release of monomers. Initially, an increase in solubilized monomers is observed with increasing digestion. As digestion continues, monomers become insoluble and precipitate from solution. HMG containing monomers, however, are more soluble under these conditions, and they become enriched at late times in digestion (Kootstra, 1982).

Together, these results suggest that micrococcal nuclease might attack transcribed genes about three times more rapidly than bulk chromatin, generating a population of digestion products slightly enriched in monomers. The solubility of active monomers might then be dramatically increased in the presence of divalent metal ions when compared to bulk nucleosomes. Specific proteins or other features of active nucleosomes might account for their increased solubility. Under other ionic conditions active nucleosomes might remain in an insoluble fraction (Davis et al., 1983). These examples emphasize that solubility differences between various nucleosome classes can have important consequences in experiments seeking to determine the concentrations of specific solubilized sequences.

DNase I Sensitivity of Active Nucleosomes

Examination of DNAs and proteins associated with monomers released by micrococcal nuclease digestion in the presence of divalent metal ions shows a correlation between enrichment of actively transcribed sequences and enrichment of nucleosomes containing HMGs 14 and 17. Are monomers containing actively transcribed sequences sensitive to digestion by DNase I, as they are in nuclei? If so, do HMGs 14 and 17 confer this sensitivity? Several reports clearly show that under appropriate conditions, monomers containing active sequences are sensitive to digestion by DNase I (Weisbrod and Weintraub, 1979). When washed in 0.35M NaCl (which quantitatively strips HMGs 14 and 17), sensitivity is abolished. Sensitivity can be restored by reconstitution with purified HMGs 14 and 17 (Weisbrod and Weintraub, 1979; Weisbrod et al., 1980). These results show that DNase I sensitivity is a feature retained in isolated HMG associated monomers. HMG columns can also specifically bind active nucleosomes previously stripped of their HMGs by salt washing. Weisbrod and Weintraub (1981) determined regions of the actively transcribed chicken α-globin domain that could bind to an HMG column. The sequences contained in the bound (active) fraction precisely span the known transcription domain (determined in Weintraub et al., 1981). These results suggest that HMGs 14 and 17 account for the high level of DNase I sensitivity observed in transcribed regions, not for the intermediate sensitivity of flanking regions. Since HMGs can interact with all nucleosomes, additional features of active nucleosomes must allow them to specifically interact with HMGs 14 and 17 (Weisbrod and Weintraub, 1979). These additional characteristics are unknown. Confirmation that HMGs confer

DNase I sensitivity to isolated nucleosomes and chromatin has been provided by Gazit et al. (1980) and Sandeen et al. (1980).

Several other results, however, suggest that additional features might contribute to DNase I sensitivity. For example, Garel and Axel (1976) were unable to show DNase I sensitivity of ovalbumin sequences in monomers isolated from oviduct nuclei digested with micrococcal nuclease at 37°. This observation prompted Senear and Palmiter (1981) to determine if a sensitive structure was retained in larger oligomers. They confirmed Garel and Axel's results for digestions conducted at 37°. They found, however, that if digestions were conducted at 0° (conditions that minimize linker trimming), monomers retained some (less than half) of the DNase I sensitivity observed in nuclei. Sensitivity was independent of oligomer size. Salt elution of monomers resulted in minimal decrements in DNase I sensitivity. In addition, salt washing of *nuclei* at ionic strengths between 0.4-0.6 M removed some, but not all, DNase I sensitivity. Nicolas et al. (1983) also reported that isolated monomers, enriched in active sequences, were not particularly sensitive to DNase I. Salt-stripping had no effect, nor did reconstitution with HMGs 14 and 17. Salt elution of nuclei was also found to have minimal effects on the DNase I sensitivity of active genes. Other workers, analyzing the DNase I sensitivity and gene content of nucleoprotein particles containing HMGs 14 and 17, have concluded that these particles are not particularly enriched in transcribed sequences (Levinger et al., 1981; Reudelhuber et al., 1982; Nicolas et al., 1983), and are not preferentially sensitive to DNase I (Nicolas et al., 1983). Unfortunately, in these experiments it is difficult to rule out nonspecific rearrangements of HMGs during preparative procedures.

While HMGs 14 and 17 or other salt elutable factors might play an important role in establishing DNase I sensitivity, the magnitude of their role does not seem to be unquestionably established. It appears that, in vivo, multiple features contribute to DNase I sensitivity.

Nucleosome Structure of Active Genes

It has been implicit in this discussion that at least a portion of actively transcribed genes are organized into canonical nucleosomes. Several experiments support this assumption. Liquid hybridization of liver cDNAs to DNA purified from mononucleosome-sized particles generated by micrococcal nuclease digestion of rat liver nuclei shows that transcribed liver sequences are as abundant in the monomer population as in total DNA. Titration with globin cDNA probes also shows that the concentration of globin genes in DNA prepared from monomers obtained from erythrocytes is identical to their concentration in total erythrocyte DNA (Lacy and Axel, 1975). Similar results were obtained by others (Bellard et al., 1977). While it is not clear from these results what proportion of the total cellular content of active genes is included in the monomer population, it seems certain that a high proportion of active gene sequences can be found in structures resembling canonical nucleosomes.

While a proportion of actively transcribed genes may be organized into a structure similar to that of bulk DNA, many compositional differences between active and bulk nucleosomes have also been characterized. Weisbrod (1982b) using HMG affinity columns to advantage, obtained nucleosomes enriched in active sequences. He then characterized features of the isolated nucleoprotein complex. It should be reiterated that the basis for the selective binding of active monomers to HMG columns is unknown. Swerdlow and Varshavsky (1983) have noted that HMGs have a dramatically increased affinity for nucleosomes possessing slightly longer lengths of DNA, which might account for the binding. However, it is not certain that additional whiskers of DNA play any role in binding active monomers to the column, and data suggesting that it does not has been reported (Weisbrod, 1982b). Regardless of the basis of binding, active monomers were found to contain DNA depleted in 5-methyl-cytosine, a modified base associated with transcriptional inactivity. In addition, Ball et al. (1983) have shown that 5-methyl-C containing DNA is enriched in nucleosomes containing H1. Together, these results suggest that active regions are depleted in H1. Nucleosomes bound to the HMG column also have an altered conformation that allows H3 sulfhydryls to be readily oxidized (Weisbrod, 1982b). H3 sulfhydryls in active nucleosomes are also known to be accessible in vivo to the fluorescent sulfhydryl reagent iodoacetamidofluorescein (Prior et al., 1983). Weisbrod (1982b) also found that active nucleosomes are enriched in hyperacetylated forms of H3 and H4, and that topoisomerase I activity copurified with the active fraction. Topoisomerase I is also known to be stimulated by and form tight complexes with HMGs 14 and 17 and histone H1 (Javaherian and Liu, 1983). Other investigators have found that active nucleosomes are capable of selectively binding RNA polymerase II, and may be depleted in histones H2A and H2B (Baer and Rhodes, 1983), that a selective class of active nucleosomes might be enriched in ubiquintinated-H2A (*Drosophila* heat shock genes; see Levinger and Varshavsky, 1982), or be enriched in other nonhistone proteins (ribosomal genes; see Prior et al., 1983).

In addition to these many compositional differences, several investigations suggest that nucleosome structure is disrupted or that major conformational changes occur during transcription. Electron microscope studies, for example, suggest that active ribosomal (McKnight et al., 1979; Labhart and Koller, 1982) and nonribosomal (Andersson et al., 1982) genes are devoid of nucleosomes. Recent biochemical studies of active genes also show that nucleosome disruption occurs across actively transcribed ribosomal (Johnson et al., 1979; Davis et al., 1983; Prior et al., 1983; Ness et al., 1983; Palen and Cech, 1984; Spadafora and Crippa, 1984), heat shock (Wu et al., 1979; Levy and Noll, 1981; Samal et al., 1981; Levinger and Varshavsky, 1982; Karpov et al., 1984), immunoglobulin (Pfeiffer and Zachau, 1980; Rose and Garard, 1984), ovalbumin (Bellard et al., 1982; in this last case a previous investigation using different techniques showed that the gene was organized into a nucleosomal structure; see Bellard et al., 1977), or globin genes (Cohen and Sheffery, 1985). In addition, Javaherian and Fasman (1984) have shown that nick translation of nuclei (Levitt et al., 1979) labels particles (presumably representing an active fraction) that sediment more slowly than bulk monomers (9S versus 11S). The hydrodynamic and compositional properties of particles from active genes, which might contaminate monomer populations, suggest that they represent either an altered nucleosomal conformation or are nucleoprotein species largely devoid of histones. Lastly, assembly of nucleosomes in oocytes and an analysis of polymerase II molecules associated SV40 minichromosomes suggests that the actively transcribed fraction is in a torsionally strained configuration that is essential for transcriptional activity (Ryoji and Worcel, 1984; Luchnick et al., 1982; see p. 72).

The results described previously all consistently suggest that nucleosomes or nucleoproteins containing actively transcribed genes are different from nucleosome containing inactive sequences. The exact differences are still a matter of investigation. Nevertheless, some data suggest that actively transcribed genes are organized into structures indistinguishable from bulk nucleosomes. To account for these apparent contradictions it seems reasonable to suggest that canonical nucleosomes can convert to a more extended structure comprised of a symmetrically unfolded histone octamer and specific nonhistone proteins (Weintraub et al., 1976; Prior et al., 1983). An equilibrium between a compact nucleosomal and an unfolded or nonnucleosomal structure that fluctuates in response to changes in the transcriptional rate of a gene seems to provide an attractive model that might reconcile many apparently contradictory observations. The degree of disruption might be related to transcription rate, however; and for genes transcribed at high rates histones might be completely displaced from upstream flanking and coding regions during periods of rapid transcription (Karpov et al., 1984).

Nucleosome Phasing

What role does the precise position of nucleosomes play in gene activation? Both micrococcal nuclease and methidiumpropyl-EDTA Iron(II) (Cartwright et al., 1983) cut between nucleosome linkers and, thus, using indirect end-labeling probes, it is possible to map nucleosome positions or phasing patterns. Micrococcal nuclease, however, has a high sequence specificity (Dingwall et al., 1981; Horz and Altenburger, 1981), and it is imperative to include naked DNA controls in phasing studies.

Globally, it has been reported that center to center distances between nucleosomes are those expected on a statistical basis (Strauss and Prunell, 1982). Several reports show, however, that under some circumstances nucleosomes might be phased or aligned, and that their position might play an important role in regulating gene expression. Nucleosomes are aligned, for example, across intergenic regions of *Drosophila* histone genes as assayed by both micrococcal nuclease (Samal et al., 1981; Worcel et al., 1983) and methidiumpropyl-EDTA iron(II) (Cartwright et al., 1983). Nucleosome position might also be crucial for the activation of tRNA genes (Wittig and Wittig, 1982). In *Xenopus laevis* tRNA genes, however, nucleosome phasing is established only under conditions of transcriptional inactivity and chromatin condensation, suggesting that phasing occurs under conditions of higher order packing (Bryan et al., 1981). In this regard it is worth noting that micrococcal nuclease digestions of *naked* DNA have shown a periodic distribution of sensitive sites across intergenic regions, but not exons (Keene and Elgin, 1981). These intergenic sequences might define regions of preferential nucleosome position that could aid packing of inactive genes in higher order structures during chromosome condensation. Exons, under selective pressures to preserve protein coding information, might thus be constrained from exhibiting this periodic structure.

Methylation of Active Regions

Alteration of the DNA template is an appealing way to mark and propagate specific chromatin structures (Holliday and Pugh, 1975). One template alteration that is related to gene expression is DNA methylation. A convenient way to assay gene methylation is to cleave purified DNA with methyl-sensitive restriction endonucleases (Bird and Southern, 1978; Waalwijk and Flavell, 1978). Most eukaryotic DNA is methylated at cytosines contained in CpG dinucleotides (Razin and Riggs, 1980). Restriction enzymes with CpG in their recognition sequence (and sensitive to cytosine methylation at this position) are therefore useful probes for assaying site specific DNA methylation patterns. A partial list of such enzymes includes Hpa II, Msp I, Hha I, Ava I, FnuD II (Tha I), Xho I, Sst II, and Sma I.

Among the first reports to show tissue specific differences in gene methylation patterns was McGhee and Ginder's demonstration (1979) that the chick globin genes were hypomethylated in erythrocytes. Since then, generally good correlations have been observed between hypomethylation of DNA and gene activation. Figure 2-3 shows, for example, that the murine α1-globin gene in mouse kidney and liver or in 3T3 cells is more methylated than in uninduced MEL cells (which are committed to α1-globin gene expression). While sites upstream of the α1-globin gene are fully unmethylated, other sites within the α1- and Bmaj-globin gene transcriptional domains are fully methylated in uninduced or induced MEL cells. Although novel upstream DNase I hypersensitive sites are established during transcriptional activation, the methylation pattern does not change. These results suggest that if hypomethylation of the globin genes is required for activation, at least some of these changes might have already occurred in MEL cells (Sheffery et al., 1982). Many other reports have shown a strong correlation between DNA hypomethylation and tissue specific gene activation. Excellent correlations were described for the chick α-globin gene domain (Weintraub et al., 1981). Virtually every Hpa II site within the transcribed domain is unmethylated

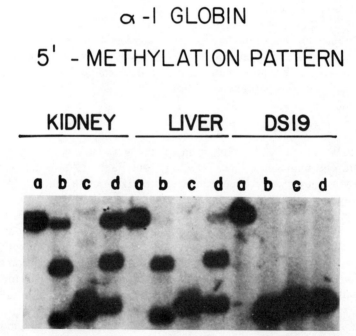

Figure 2-3. Methylation of the α1-globin gene in mouse tissues and MEL cells (line DS 19). DNAs prepared from mouse kidney and liver, or from MEL cells were cleaved with Sac I (a) and Sac I plus Hha I (b), Msp I (c), or Hpa II (d). Digests were separated by electrophoresis, blotted to nitrocellulose, and hybridized to probe αZ (see Fig. 2-1A).

1982; in vitro studies are to be discussed). However, the relationships can be complex. For example, sites upstream of the $\alpha 2(I)$ collagen gene are unmethylated regardless of whether the gene is expressed (McKeon et al., 1982). Other upstream regions that are unmethylated regardless of expression have been reported (the hamster aprt and mouse dhfr genes; Stein et al., 1983). The downstream region of the $\alpha 2(I)$ collagen gene is heavily methylated regardless of its expression, and downstream methylation does not interfere with the relative DNase I sensitivity of this region in actively expressing cells. Changes in DNase I hypersensitive sites at the promoter also occur without apparent changes in methylation patterns (McKeon et al., 1984a). Heavy methylation of the transcribed region of *X. laevis* rDNA is also compatible with DNase I sensitivity (Macleod and Bird, 1982). Other reports have shown that extinction of albumin gene expression can occur without methylation (Ott et al., 1984), that developmental gene activation can be correlated with an *increase* in gene methylation (Tanaka et al., 1983), and that insulin gene methylation is not influential in regulating its expression (Cate et al., 1983). Finally, in two studies it has been possible to compare the timing of gene demethylation to transcriptional activation. Vitellogenin gene transcription in chicken livers is inducible with steroids. Transcriptional activation is associated with the loss of at least one methylated C upstream of the gene (Wilks et al., 1982; Burch and Weintraub, 1983). The downstream region remains highly methylated and resembles the pattern found in nonexpressing cells (Folger et al., 1983). Demethylation of the upstream region occurs after transcriptional activation (Burch and Weintraub, 1983), suggesting that transcription might play a role in blocking or altering methylation, instead of demethylation allowing transcription. The hypomethylation of genes encoding δ-crystallin also appear to demethylate *after* δ-crystallin production has been initiated (Grainger et al., 1983).

Despite these complexities, in vitro assays strongly suggest that methylation of DNA substrates can directly influence gene transcription. Although DNA methylation does not affect in vitro transcription reactions (Doerfler, 1983), many in vitro methylated genes are not transcribed after microinjection into *Xenopus* oocytes (Vardimon et al., 1982; Fradin et al., 1982) or transfection into L-cells (Stein et al., 1982). This effect is not general, since methylated, microinjected *Xenopus laevis* rDNA is efficiently transcribed (Macleod and Bird, 1983). In vitro experiments have clearly demonstrated, however, that methylation of a few specific sequences near promoter regions can have a dramatic effect on gene transcription when assayed in either transient or long-term transfection experiments. Thus, methylation of select Hpa II or Hha I sites upstream from the Ad 12 protein IX promoter severely

during a developmental transition that includes increased DN
DNase I hypersensitive site formation, and de novo gene tra

Interestingly, experimental demethylation of transcription
can sometimes result in gene activation. Inactive methylate
have been reactivated by cloning, which eliminates eukaryotic
patterns (Harbers et al., 1981; Simon et al., 1983; Jahner
Treatment of host cells with 5-aza-cytidine, a cytosine analog th
methylated at the 5-position and also potently inhibits DNA methyl
when incorporated into DNA (Jones and Taylor, 1981; Creusot
Christman et al., 1980), can also induce previously silent endoge
genes (Groudine et al., 1981b). In this latter case 5-aza-C was able
transcription of an inactive viral gene, but it did not reactivate
criptionally inert globin genes in the same cells. Thus the effects o
can be selective. Expression of the metallothionein genes has a
induced by 5-aza-C coordinately with their DNA demethylation (C
and Palmiter, 1981). 5-aza-C has also been shown to induce novel diff
tion programs in fibroblasts (Jones and Taylor, 1980), and to increa
hemoglobin producing cells (F-cells) in a B thalassemic patient (Ley
1982). However, interpretation of these latter results must be cautious
5-aza-C is toxic, and its mechanism of gene activation is not unders
Indeed, hydroxyurea, a DNA synthesis inhibitor, also increases circul
F-cells in thalassemics (Letvin et al., 1984). In addition, treatments that
induce DNA repair systems (UV irradiation) also induce gene demethyla
and activation (Lieberman et al., 1983).

Other in vivo results suggest complex relationships between DNa methy
tion and gene activation (Bird, 1984). Before reviewing some of these resul
however, it must be understood that many observations rely on sampling th
methylation state of only the few CpG dinucleotides accessible to restrictio
endonucleases. A method capable of assaying the methylation status o
every CpG dinucleotide along a DNA sequence has been developed (Church
and Gilbert, 1984). It relies on assaying the altered reactivity of 5-methyl-C to
Maxam-Gilbert sequencing reagents. Reactions are conducted on total genomic
(i.e., uncloned) DNA and analyzed by blot hybridization and indirect end-
labeling methods. It must be kept in mind, therefore, that some conclusions
regarding site-specific DNA methylation and gene control to be cited may
have to be modified when the newer technique can be applied. Even so,
several in vivo results using only restriction endonucleases suggest that
demethylation at only a few specific sites can be correlated with gene
expression (van der Ploeg and Flavell, 1980; Shen and Maniatis, 1980). Other
results also suggest that methylation of discrete upstream regions can play
critical roles in regulating gene expression (in vivo examples: hamster aprt
and mouse dhfr gene, Stein et al., 1983; mouse albumin genes, Ott et al.,

affects transcription from the promoter in transient assays (Kruczek and Doerfler, 1983). Busslinger et al. (1983) also showed that methylation of nucleotides from −760 to +100 eliminated transcription of γ-globin gene sequences stably transfected into L-cells. Methylation of regions further upstream or throughout the body of the gene had little or no effect on expression. These results strongly suggest that methylation of promoter proximal sequences dramatically and directly affects gene expression under certain conditions.

Finally, other in vitro assays suggest that DNA methylation might directly influence chromatin structure as well as promoter function. The presence of 5-methyl-cytosine in poly (dCpG) permits the formation of Z-DNA under nearly physiological conditions (Behe and Felsenfeld, 1981). When DNA is in the Z conformation, nucleosomes are not formed readily (Nickol et al., 1982). Whether these potential structural transitions play a role in modifying chromatin structure in vivo is unknown.

HYPERSENSITIVE SITES, ACTIVE CHROMATIN, AND CHANGES IN CHROMATIN STRUCTURE DURING DEVELOPMENT, DIFFERENTIATION, AND GENE ACTIVATION

DNase I Hypersensitive Sites

DNase I hypersensitive sites, alluded to several times above, are small chromatin regions exceedingly sensitive to attack by DNase I and a variety of other nucleases and chemicals. A conventional method for mapping DNase I hypersensitive sites is via indirect end-labeling of restriction fragments (Wu, 1980; Nedospasov and Georgiev, 1980). Nuclei are treated briefly with DNase I or other reagents, DNA is purified, cleaved with a restriction enzyme, and transferred to nitrocellulose after electrophoresis. A short, cloned probe that abuts one end of the restriction fragment is used in the hybridization reaction. The length of visualized subfragments generated by nuclease cleavage maps the distance of the nuclease cleavage site from the restriction site. An example of using the indirect end-labeling procedure to map nuclease hypersensitive sites near the murine α1-globin gene in MEL cells is shown in Figure 2-4. A weak DNase I hypersensitive site, mapping about 200 bp upstream of the cap site marks the α1-globin gene before transcription. After induction, the sensitivity of the region increases, and the hypersensitive site is now resolved as a doublet—perhaps more aptly described

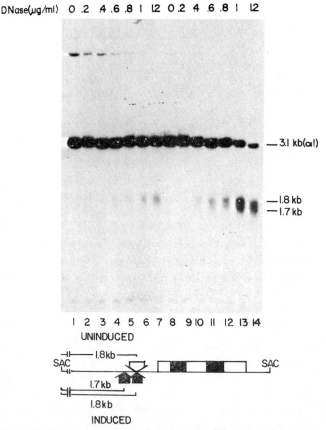

Figure 2-4. DNase I hypersensitive sites near the α1-globin gene. Nuclei prepared from uninduced MEL cells (left) or cells induced by culture in the presence of 5 mM hexamethylenebisacetamide (HMBA; right) were incubated with the amounts of DNase I indicated above the lanes. Purified DNA was cleaved with Sac I, separated by electrophoresis, blotted to nitrocellulose, and hybridized to probe αZ (Fig. 2-1A). A 3.1 kb Sac I fragment containing the α1-globin gene is indicated, as are DNase I generated subfragments of about 1.8 kb (detected in both uninduced and induced cells) and 1.7 kb (detected only in induced cells). Locations of cleavage sites are indicated diagrammatically by vertical arrows at the bottom of the Figure. Open arrow: cleavage site detected in uninduced cells; filled arrows: cleavage sites detected after induction. *(From Sheffery et al., 1984, p. 425; copyright © 1984 by Academic Press Inc. [London] Ltd.)*

as a broad region with a node (see Wu, 1984). Identical mapping information can be generated without using indirect end-labeling probes by comparing lengths of subfragments generated in several restriction digests to a known restriction map.

Sites hypersensitive to digestion by DNase I have been demonstrated upstream of almost every active gene examined. Sites typically extend 200 or more bp upstream of the cap site, sometimes including, sometimes excluding the TATA box (recent reviews have also appeared: see Elgin, 1981, 1984). A partial list of examples includes active chick α- and β-globin genes (Stalder et al., 1980b; Weintraub et al., 1981; McGhee et al., 1981), murine α- and β-globin genes (Sheffery et al., 1982, 1984; Hofer et al., 1982; Balcarek and McMorris, 1983; Smith and Yu, 1984), human β- and Σ-globin genes (Groudine et al., 1983; Tuan and London, 1984), chick ovalbumin (Kaye et al., 1984), conalbumin (Kuo et al., 1979), $\alpha 2(I)$ collagen (McKeon et al., 1984a), vitellogenin (Burch and Weintraub, 1983), and lysozyme (Fritton et al., 1983) genes, murine and human immunoglobulin genes (Weischet et al., 1982; Chung et al., 1983; Parslow and Granner, 1983a, 1983b; Mills et al., 1983; Picard and Schaffner, 1984), *Drosophila* heat shock (Wu, 1980; Keene et al., 1981; Samal et al., 1981), glue (Shermoen and Beckendorf, 1982; McGinnis et al., 1983a, 1983b), and ribosomal protein genes (Wong et al., 1981), rat insulin genes (Wu and Gilbert, 1981), sea urchin histone genes (Bryan et al., 1983), yeast mating type (Nasmyth, 1982) and acid phosphatase (Bergman and Kramer, 1983) genes, *tetrahymena* ribosomal genes (Borchsenius et al., 1981; Bonven and Westergaard, 1982; Palen and Cech, 1984), a transfected herpes thymidine kinase gene (Sweet et al., 1982), and a variety of viral promoter regions (Scott and Wigmore, 1978; Varshavsky et al., 1979; Saragosti et al., 1980, 1982; Jakobovits et al., 1980; Herbomel et al., 1981; Groudine et al., 1981b; Schubach and Groudine, 1984).

DNase I hypersensitive sites appear ubiquitously associated with active or "primed," potentially active genes (Wu, 1980; Burch and Weintraub, 1983; Sheffery et al., 1984). Sites most frequently remarked on are located within 1 kb upstream of a cap site. However, hypersensitive sites located as far upstream as 4-6 kb have been described (ovalbumin: Kaye et al., 1984; Σ-globin: Tuan and London, 1984). In addition, sites have been reported downstream of transcription units (chick globins: Weintraub et al., 1981; chicken vitellogenin: Burch and Weintraub, 1983), within introns (mouse globins: Hofer et al., 1982; Sheffery et al., 1983; Balcarek and McMorris, 1983; human c-myc: Tuan and London, 1984; chicken c-myc: Schubach and Groudine, 1984; and in immunoglobulins, see previous references), and within exons (chick globins: Weintraub et al., 1981; *Drosophila* glue: Shermoen and Beckendorf, 1982; c-myc: Tuan and London, 1984; Schubach and Groudine, 1984).

Functional Significance of DNase I Hypersensitive Sites

The most compelling evidence that hypersensitive sites have an important function in gene expression is afforded by a combination of genetic and biochemical data available in *Drosophila*. Shermoen and Beckendorf (1982) analyzed a complex of five DNase I hypersensitive sites located at +30, −70, −330, −405, and −480 from the active *Drosophila* glue protein Sgs4. This locus was chosen because numerous mutants had been identified that make little or no glue protein. In one mutant, a small deletion removed sequences corresponding to the −330 site. This strain showed a fiftyfold reduction in gene expression, but retained all hypersensitive sites (except the −330 site) at their normal sequences (i.e., they did not form a fixed distance from the cap site, but over specific DNA sequences). A second deletion removed sequences corresponding to the −405 and −480 sites. This mutant, a null, had none of the hypersensitive sites detected in wild-type flies, even though their DNA sequences were presumably intact. Together, these results suggest that hypersensitive sites have a crucial function in establishing gene expression. Studies of additional mutants showed that point mutations within known hypersensitive regions did not prevent hypersensitive site formation, but reduced gene expression by 50%. In another strain, scattered point mutations prevented hypersensitive site formation and dramatically decreased gene expression (McGinnis et al., 1983*a*). Finally, when insertion of a transposable element displaces the far upstream hypersensitive sites 1.3 kb further upstream from the Sgs4 promoter, DNase I hypersensitive sites still form over the displaced sequences, although expression of the gene is affected, presumably by the control of transcript initiation from the transposable element (McGinnis et al., 1983*b*).

Other functions have been attributed to hypersensitive sites. For example, it is known that origins of replication can be hypersensitive to DNase I (Saragosti et al., 1982; Palen and Cech, 1984), but it is unlikely that all hypersensitive sites are origins of replication. Some hypersensitive sites map far upstream of normal transcription start sites (Kaye et al., 1984; Tuan and London, 1984). Of the six hypersensitive sites mapping as far as −6 kb from the human Σ-globin cap site (Tuan and London, 1984), four appear to be associated with far-upstream transcription starts described by Allan et al. (1983). Some of these starts might be transcribed by pol III (Carlson and Ross, 1983). It therefore seems possible that some hypersensitive sites might be correlated with accessibility to pol III or might define pol III promoters. Hypersensitive sites have also been found in regions where yeast mating-type switching occurs (Nasmyth, 1982), where heavy chain switching can occur in B lymphocytes (Storb et al., 1979), and in regions where avian leukosis virus integrates near the c-myc gene in chickens (Schubach and Groudine, 1984). Hypersensitive sites might thus be targets for DNA rearrangements in eukaryotes. Finally, there is evidence that enhancer elements and hypersensitive sites have many common features (see following).

The Fine Structure of Hypersensitive Sites and Their Relationship to Enhancers

DNase I has been extensively used to map hypersensitive sites, but these regions are accessible to a variety of double- and single-strand specific nucleases including DNase II, micrococcal nuclease, restriction endonucleases, S1 nuclease, Neurospora crassa nuclease, and Bal 31. In addition, chemical reagents that make double-strand breaks in duplex DNA, such as 1,10 phenanthroline-cuprous complex (Cartwright et al., 1982; Jessee et al., 1982), and methidiumpropyl-EDTA iron (II) (Cartwright et al., 1983), or reagents that form adducts with unpaired adenines and cytosines, such as bromoacetaldehyde (Kohwi-Shigematsu et al., 1983) also specifically react with hypersensitive sites. The putative single-strand specific reactivities of some of these reagents suggest that hypersensitive sites are in an unusual conformation that may play an important role in their action. The sensitive sites for these various reagents do not necessarily overlap. Figure 2-5, for example, shows the S1 nuclease sensitive regions upstream of the murine α1-globin gene in MEL cells. An S1 site (which overlaps with the comparable DNase I site detected in uninduced cells; see Fig. 2-4) marks the gene before transcription. After induction, a novel S1 nuclease site appears mapping *away* from the novel DNase I site, and close to the cap site (compare with Fig. 2-4). These results show that hypersensitive sites can have heterogeneous domains of nuclease sensitivity.

A fine structure analysis of nuclease sensitive regions upstream of other genes confirms this view. McGhee et al. (1981) found that a region extending from -65 (upstream of the TATA box, but including the CCAAT box) to -260 was accessible to DNase I, DNase II, micrococcal nuclease, and restriction endonucleases. Some sites did not overlap. In particular, a micrococcal nuclease site in the region -230 to -260 was distinct from the DNase I region (-65 to -160). A hint that hypersensitive regions may be partially protein free was provided by cleaving nuclei with Msp I. This restriction endonuclease released 50-80% of the total DNA between two Msp I sites at -109 and -224. About 35% of the released DNA (that is about 20-30% of the total) behaved under electrophoretic and blot transfer conditions as protein free DNA. This is reminiscent of the hypersensitive site in the SV40 minichromosome, a region that appears devoid of nucleosomes in the electron microscope (Jakobovits et al., 1980; Saragosti et al., 1980).

Burch (1984) also fine mapped four hypersensitive sites upstream of the chicken vitellogenin gene. These sites range from +27 to −960. Sites could be mapped precisely (to within 20 bp) by comparing nuclease cleavage sites to several restriction endonuclease sites. Comparison of the sensitive sites to

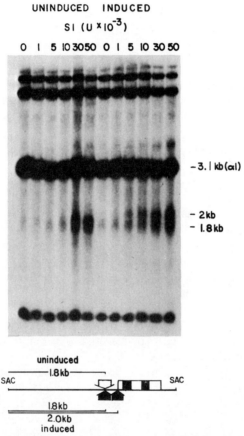

Figure 2-5. S1 nuclease sensitivity of chromatin containing the α-globin gene. Nuclei from uninduced (left) or induced (right) MEL cells were incubated with the amounts of S1 nuclease indicated above each lane. DNA was prepared, cleaved with Sac I, separated by electrophoresis, blotted to nitrocellulose, and hybridized to probe αZ (Fig. 2-1A). The 3.1 kb Sac I fragment containing the α1-globin gene, and subfragments of 1.8 kb and 2.0 kb generated by S1 nuclease are indicated. Locations of S1 nuclease cleavage sites are indicated at the bottom of the figure by vertical arrows. Open arrow: S1 cleavage site detected in uninduced cells; filled arrows: cleavage sites detected after induction. *(From Sheffery et al., 1984, p. 427; copyright © 1984 by Academic Press Inc. [London] Ltd.)*

the underlying DNA sequence revealed no correlation between nuclease hypersensitivity and A-T rich regions (a similar conclusion was drawn by McGhee et al., 1981), and no strong correlation to regions of inverted or direct repeats or potential Z-forming sequences (i.e., alternating purine/pyrimidine tracts). However, one site (extending to −960) contained an 11 bp region identical to the central 9 bp of a proposed enhancer core sequence (Khoury and Gruss, 1983).

The association of hypersensitive sites with enhancer elements had already been established by a fine mapping study of hypersensitive sites in the SV40 minichromosome (Saragosti et al., 1982). The SV40 hypersensitive site, spanning the origin of replication and the early and late promoters, is comprised of a sensitive region of about 390 bp. Within the sensitive region, a resistant node can be detected that spans the 21 bp repeats. Six hypersensitive sites map within the sensitive domain. Three map near T-antigen binding sites (Tjian, 1978), and one of these maps to the origin of replication. Three other sites mapped into the late region. Of these three sites, two mapped within the 72 bp-repeat enhancer element. The sensitivities of these sites were not affected by washing with 0.35 M salt, suggesting that HMGs 14 and 17 play no role in their maintenance. That two hypersensitive sites were located within the enhancer element and others were located nearby suggests that one role of enhancers might be to organize hypersensitive sites in chromatin. This idea has received additional support from the observation that the tissue-specific immunoglobulin enhancers located within introns (Queen and Baltimore, 1983; Gillies et al., 1983; Banerji et al., 1983) are also hypersensitive sites (Weischet et al., 1982; Chung et al., 1983; Parslow and Granner, 1983a, 1983b; Mills et al., 1983; Picard and Schaffner, 1984). In addition, movement of the SV40 enhancer element to new sequence environments induces the formation of new nucleosome free regions that are hypersensitive to DNase I (Fromm and Berg, 1983; Jongstra et al., 1984). Several known hypersensitive sites have homologies to enhancer core sequences. In addition to the chicken vitellogenin gene cited above, a site at −6 kb from the human Σ-globin gene has also been reported to have unusual features that include a stretch of 28 consecutive pyrimidines (Ts), a long (21 bp) region of alternating purines and pyrimidines, and three regions showing homologies to enhancer core sequences (Tuan and London, 1984).

These many similarities suggest that hypersensitive sites and enhancer elements share some common function. Similar functions might include sequestering of transcription factors essential for differentiation-specific gene function, or establishing attachment sites on the nuclear matrix that might be required for transcription. The understanding of these similarities and their relationship to gene function awaits future experimentation.

Unusual DNA Conformations in Nuclease Hypersensitive Sites

The DNA sequences of the SV40 enhancer element can induce local changes in chromatin structure. These changes are manifest by an exclusion of normal nucleosomes and creation of nuclease sensitive sites. In addition, hypersensitive sites have unusual conformations that make them sensitive to S1 nuclease and to chemical reagents known to interact with unpaired bases (Larsen and Weintraub, 1982; Kohwi-Shigematsu et al., 1983; see also Fig. 2-5). These results suggest that DNA sequence information itself might play a critical role in establishing features of hypersensitive sites. This possibility was tested directly by Weintraub (1983), who showed that transfection of the chicken B_{maj}-globin gene promoter region into L-cells established a structure that was sensitive to S1 nuclease (endogenous globin genes were resistant). Similar S1 nuclease sites were detected in supercoiled plasmids in vitro, and these regions were found to bind nucleosomes poorly in in vitro reconstitution assays. These results led many investigators to map S1 nuclease sensitive sites in chromatin or in supercoiled plasmids (Nickol and Felsenfeld, 1983; Schon et al., 1983; Shen, 1983; Cockerill and Goodwin, 1983; Mace et al., 1983; McKeon et al., 1984b; Finer et al., 1984). Many experiments using supercoiled plasmids in vitro showed that S1 nuclease cleaved in stretches of polypurine/polypyrimidine tracts. An example of the S1 nuclease cleavage pattern near the murine α1-globin promoter region is shown in Figure 2-6A. Although the DNA sequence of the α1-globin gene contains a 19 bp inverted repeat sequence capable of forming a stable cruciform structure, S1 cleaves in a long (over 100 bp) uninemic stretch of polypurine/polypyrimidine (there are six pyrimidines in a tract containing over 100 purines; Fig. 2-6B). Cleavage is dependent on the superhelical state of the plasmid, and the S1 cleavage sites detected in plasmids do not seem to overlap the cleavage sites detected in chromatin.

These results suggest that hypersensitive sites can form unusual conformations in vivo. Some of these features might be mimicked in vitro if specific DNA sequences were under torsional stress. Torsional stress might have important consequences for the regulation of eukaryotic genes. For example, if promoter regions of active genes were under torsional stress a class of regulatory proteins might exist to take advantage of the structural changes that might be induced. On a simpler level, negative supercoils make the energetics of strand untwisting a more favorable process, and this feature might be important for processes that initiate transcription. Recall that bulk DNA is wrapped around histone octamers, and the coiling of the DNA axis about nucleosomes is, of course, equivalent to supercoiling the duplex. However, most internucleosomal DNA is not under torsional stress, since

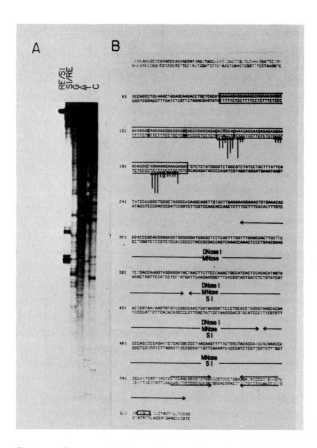

Figure 2-6. Cleavage of supercoiled plasmids containing the α1-globin promoter region by S1 nuclease. *(A)* Supercoiled plasmids containing αA (see Fig. 2-1A), cloned in plasmid pUC9, were purified and digested Eco RI followed by S1 nuclease (S1/RE), or with S1 nuclease (in 30 mM NaOAc, pH 4.6, 50 mM NaCl, 3 mM ZnSO4) followed by Eco RI (S1/RE). Digests were boiled and hybridized to the M13 "reverse" sequencing primer. Primer extension was conducted in the presence of all four deoxyribonucleotide triphosphates (RE/S1, S1/RE) or in the presence of one of the dideoxyribonucleotide triphosphates indicated at the top of the lanes (GATC). *(B)* Nucleotide sequence of fragment αA (derived from sequencing gels similar to that shown in panel *(A)* showing S1 nuclease cleavage sites detected in the αA promoter region in supercoiled plasmids. A long stretch of polypurine/polypyrimidine is boxed; pyrimidines in the largely purine strand (and vice versa) are indicated by heavy lettering. S1 nuclease cleavage sites are indicated by vertical arrows below the strand cleaved. The length of the arrow is proportional to the intensity of the band produced by cleavage. A region near the TATA box containing an inverted repeat sequence is indicated by arrows. Approximate boundaries of regions cleaved in nuclei by DNase I, S1 nuclease, and Micrococcal nuclease are also indicated.

SV40 minichromosomes (DNA nucleosome complexes) appear to be mostly untwisted, relaxed molecules (Germond et al., 1975), and bulk nuclear chromatin is not torsionally stressed (Sinden et al., 1980). When histones are removed from SV40 minichromosomes by protein extraction, the superhelicity of the molecule formerly constrained to DNA associated with the nucleosome core is released and the purified DNA is isolated as a twisted, supercoiled structure (Germond et al., 1975). Thus, one simple way to place an active region under torsional stress would be to release the superhelicity of DNA constrained to nucleosome cores into the internucleosomal DNA. This could be achieved by specifically removing nucleosomes spanning active domains, or by altering nucleosome conformation so that torsion was not constrained to the nucleosome and could be transmitted to adjacent sequences. Note that if active chromatin regions were in a torsionally strained configuration, nicking would cause the active domain to relax in an all or none pattern. Nicking internucleosomal linker regions spanning relaxed domains, such as those that exist between bulk nucleosomes, would, of course, have no comparable effect.

The above analysis leaves us with an experimental question: are hypersensitive sites or active chromatin regions torsionally strained? Some experimental evidence supports the view that they are. First, Luchnick et al. (1982) found that 2-5% of SV40 minichromosomes behaved as though they were under torsional stress as assayed by their response to topoisomerase I. Torsionally stressed minichromosomes were enriched (20% of all molecules) in regions of sucrose gradients containing active transcription complexes. The process of topoisomerase I induced relaxation resembled a spring-loaded, unconstrained uncoiling, since it appeared to be all or none. Second, molecules microinjected into *Xenopus* oocytes are transcribed efficiently only if they are circular and, thereby, under topological constraint. Linearization of transcribing molecules in vivo results in cessation of transcription (Harland et al., 1983). However, both initiation and elongation of endogenous *Xenopus* ribosomal DNA templates is unaffected by cleavage and linearization with restriction endonucleases (Pruitt and Reeder, 1984). It is possible that, in vivo, endogenous genes are kept under torsional strain by stable, localized structures, and that supercoiling of microinjected plasmid DNAs (as they bind histones) mimics the effects of proteins bound locally in vitro. Third, using *Xenopus* oocytes in an in vivo chromatin assembly system, Ryoji and Worcel (1984) found that the actively transcribed fraction of the microinjected minichromosomes (about half of the total) was under torsional strain as assayed by its relaxation by topoisomerase I. Relaxation was found to be an all or none phenomenon; that is, unwinding was unconstrained. Since virtually all molecules were shown to be assembled into a periodic structure resembling nucleosomes, but one-half displayed unconstrained DNA supercoils that could be quantitatively relaxed by topoisomerase I, it was concluded that the active structures must be in an altered nucleosomal conformation that does not constrain unwinding as the canonical nucleosome structure might (Glikin et al., 1984). In addition, novobiocin, an inhibitor of type II topoisomerases, inhibited the formation of the active templates. Novobiocin is also an inhibitor of the prokaryotic topoisomerase, DNA gyrase. Gyrase can introduce superhelical turns into closed DNA duplexes. The results of Ryoji and Worcel thus suggested that a process mediated by a gyrase-like activity might help to establish the torsionally strained active complexes. Finally, these observations have been extended very recently by Villeponteau et al. (1984) who found that novobiocin could inhibit maintenance of DNase I sensitive domains. Thus, in the presence of the drug, the DNase I sensitivity of active domains is lost. They concluded that torsionally strained domains around active genes promote the observed DNase I sensitivity. The results of Villeponteau et al. help to explain some of the paradoxes, alluded to earlier, regarding the DNase I sensitivity of active nucleosomes. It was noted then (see p. 58) that many workers had found active nucleosomes to be no more sensitive to DNase I than bulk DNA. Villeponteau et al. (1984) explain these results by suggesting that cleavage of chromatin with micrococcal nuclease (in order to prepare mononucleosomes) releases the torisonal stress required for maintenance of the sensitive structure.

These results indicate that actively transcribed domains may be torsionally strained. The torsion might be generated by release of superhelices constrained to nucleosomes into the surrounding DNA as a result of a conformational change in nucleosome structure or by removal of nucleosomes. Such a conformational change in nucleosome structure might be revealed as a disruption in nucleosome structure in micrococcal nuclease digestion experiments, and some of the data supporting the nucleosome "disruption" of active regions was reviewed earlier (see p. 59). In addition, the results discussed above suggest that active processes, using the eukaryotic equivalent of a DNA gyrase, might also be responsible for generating or maintaining the torsionally strained, active structure. Torsional strain generated across the active domain might enhance transcription processes either by reducing the energy required for strand separation during transcription or by inducing conformational changes in regulatory regions that then allows binding of a novel class of regulatory molecules (or both).

Proteins Associated with Hypersensitive Sites

The SV40 hypersensitive site appears nucleosome free. Are hypersensitive sites protein free? Some results, cited previously (McGhee et al., 1981), suggest that regions of hypersensitive sites might be protein free. Other results suggest that hypersensitive sites might also be viewed as local openings in chromatin that allow greater access of regulatory proteins to important control regions. Wu (1984) analyzed the resistance of heat-shock gene chromatin to trimming by exonuclease III. After digestion of nuclei with nucleases to produce double-strand cuts in hypersensitive sites, ends were trimmed back with exo III (until a barrier was encountered) and made flush with S1 nuclease. Wu interpreted resistance to exo III resection as indirect evidence for protein binding. The results of these experiments strongly suggest that before heat shock a protein factor is bound near the TATA box. After heat shock an additional factor binds further upstream in a region known to be important in conferring heat inducibility on a neutral gene in transfection assays (Pelham, 1982). These factors protect small regions of DNA (22 to 28 bp, and 36 to < 68 bp, respectively).

Other investigators have asked if specific proteins are required to build hypersensitive sites. Emerson and Felsenfeld (1984) have identified a factor that can bind in vitro to sequences present in DNase I hypersensitive sites. This factor can also specify the establishment of hypersensitive structures during in vitro nucleosome assembly reactions.

The further characterization of proteins that specifically bind to hypersensitive sites and play a role in establishing these regions should provide new insights into hypersensitive site formation, actively transcribed chromatin, and possibly into enhancer function.

Hypersensitive Sites, Development, and Differentiation

It has already been mentioned that hypersensitive sites can mark genes *committed* to activation, before establishment of the high transcription rates characteristic of differentiation specific gene outputs. These preestablished sites are obviously propagated in the absence of stimuli that established them. Experimental induction and stable propagation of hypersensitive sites near the globin genes in fibroblasts treated with high salt or reversibly transformed with a temperature sensitive Rous sarcoma virus (Groudine and Weintraub, 1982) also shows that, once formed, hypersensitive sites can template themselves.

These results suggest that hypersensitive site formation might result from the completion of independent biochemical events not necessarily related to transcription per se. It might, therefore, be expected that during development and differentiation hypersensitive sites could be assembled and dismantled independently of one another, and in the absence of gene expression. This expectation is largely confirmed by experiments monitoring changes in hypersensitive sites that occur during development and differentiation.

During chick and human fetal development, fetal globin genes normally expressed at high rates are marked by hypersensitive sites (Weintraub et al., 1981; Groudine et al., 1983). In human fetal development, adult genes expressed at low or undetectable levels are also marked by hypersensitive sites. Thus, activation of human adult globin genes is not associated solely with the establishment of novel hypersensitive sites, and other events required for gene activation apparently must occur to express these genes (Groudine et al., 1983).

Switches in globin chain synthesis occurring in both chick and human erythropoietic development are accompanied by the dismantling of hypersensitive sites near fetal globin genes during establishment of the adult lineage (Weintraub et al., 1981; Groudine et al., 1981a, 1983). In the chick, developmental switching is also accompanied by an increase in fetal gene methylation and extinction of transcription.

Changes in chromatin structure that accompany the developmentally regulated shut-down of gene transcription have also been investigated in sea urchin histone genes (Bryan et al., 1983). Early in sea urchin development, a major histone gene repeat is transiently expressed. The active gene is characterized by DNase I and micrococcal nuclease hypersensitive sites upstream of the gene. Interestingly, a downstream site sensitive to micrococcal nuclease digestion in naked DNA is preferentially protected from digestion during active gene expression. This region is associated with a site important for generating the end of the histone mRNAs (Birchmeier et al., 1982). The transcription rate of the genes gradually decreases during development, but despite the decreased expression, hypersensitive sites remain. However, a switch that removes the hypersensitive sites (and exposes the downstream region of the gene) occurs within one to two cell cycles after most transcription has ceased.

The chicken vitellogenin gene in liver is specifically marked by hypersensitive sites within the body and in the downstream region of the gene *before* hormonal stimulation, and in the absence of detected gene expression. Thus, a developmentally regulated sequence of biochemical events, apparently not involving transcription per se, marks the vitellogenin gene in liver cells (Burch and Weintraub, 1983). Hormone administration leads to the establishment of three hypersensitive sites upstream of the gene. These sites appear concomitant with gene expression. However, two of these three sites are stably maintained after hormone withdrawal, and well after cessation of gene expression. One site, perhaps related to a steroid receptor complex binding domain, behaves independently and is maintained only in the presence of steroid. As in other examples of steroid inducible gene expression, the primary interaction of the gene with the steroid receptor complex results in a slow (within days) accumulation of mRNA. After hormone withdrawal, expression falls, but secondary stimulation results in a rapid (within hours) induction in gene expression. It is easy to imagine that upstream sites, induced by steroids and stably maintained in their absence, might be responsible for the rapidity of the secondary response. The genes are already marked, and perhaps have bound specific transcription factors. All that might be required to assemble an initiation complex would be interaction of the steroid receptor complex itself with the stably maintained factors.

The independent behavior of hypersensitive sites has also been demonstrated in MEL cells. As described above, both the α1- and Bmaj-globin genes are marked by hypersensitive sites before gene activation. The Bmaj-globin gene has a hypersensitive site within the second intervening sequence (IVS-2; Hofer et al., 1982; Balcarek and McMorris, 1983; Sheffery et al., 1983). With transcriptional activation of the Bmaj-globin gene the IVS-2 site disappears and is replaced by a site upstream from the cap (Sheffery et al., 1983; Balcarek and McMorris, 1983; Salditt-Georgieff et al., 1984). In a variant MEL cell line that does not respond to inducer, the IVS-2 site disappears but no upstream hypersensitive site develops (Sheffery et al., 1983), suggesting that the decay of the IVS-2 site does not depend on the establishment of the upstream sites. These results, and results cited above for sea urchin histone genes, suggest that while hypersensitive sites can be stably propagated for long periods of time, cells also have ways of dismantling these structures.

The timing of upstream hypersensitive site formation compared to transcriptional activation of the globin genes in MEL cells also suggests that these events are independent. Hypersensitive sites associated with the active transcription of the α1- and Bmaj-globin genes are established rapidly, within two cell cycles (about 24 hours; Sheffery et al., 1984; Salditt-Georgieff et al., 1984; Balcarek and McMorris, 1983; Smith and Yu, 1984). Increased transcription rates (twofold) are also detectable within two cell cycles, so the temporal separation is not complete. However, transcription peaks continue to increase for 48 to 60 hours (Sheffery et al., 1984; Salditt-Georgieff et al., 1984). It thus appears that in MEL cells the changes in chromatin structure slightly precede transcription peaks.

How might hypersensitive sites be established during development? Specific factors implicated in hypersensitive site formation in vitro were shown to assemble hypersensitive sites only if they were present *during* nucleosome assembly; sites were not formed if factors were added after nucleosomes formed (Emerson and Felsenfeld, 1984). One possible time for new nucleosome deposition and chromatin restructuring is DNA replication. Indeed, globin gene switching during chick development can be delayed by treating embryos with FUdR (Groudine and Weintraub, 1981). Possible *trans* acting mechanisms have also been suggested to account for the changes in chromatin structure associated with heat-shock gene activation in *Drosophila*.

These models propose that activation might partially result from selective recognition of modified (ubiquitinated) nucleosomes by enzymes activated by the heat-shock response (Levinger and Varshavsky, 1982).

In summary, nuclease hypersensitive sites have a heterogeneous internal structure with regions sensitive to one nuclease but not to another. They also display nodes of sensitivity within them that may be related to protein binding sites. Sensitive sites have been predominantly mapped within regions −200 to −400 bp of the cap site, but sites downstream from transcription domains, within exons and introns and far upstream (−6 kb) from cap sites have been described. Genetic approaches to understanding function show clearly that hypersensitive sites play an important role in establishing gene expression. In this regard hypersensitive sites resemble enhancer elements. Enhancers resemble hypersensitive sites in other ways: they can establish hypersensitive sites and regions devoid of nucleosome structures. Cellular hypersensitive sites also form structures in vitro that appear to exclude nucleosomes and are sensitive to exogenous nucleases. Hypersensitive sites may mark origins of replication and might be targets for insertion of genetic information. Hypersensitive sites are assembled and disassembled in an orderly fashion during development. After formation, sites can template themselves in the absence of signals that initiated their formation. Most of the evidence suggest that hypersensitive site formation and interaction represent *independent* biochemical events, and that transcription per se is not likely to be responsible for their establishment. How hypersensitive sites are established in vivo remains an open question, but replication in the presence of specific protein factors may be involved in chromatin restructuring.

ACKNOWLEDGEMENTS

This work was supported, in part, by grants CA-31768 and CA-08748 from the National Cancer Institute, the Bristol-Meyers Foundation, and the Charles H. Revson Foundation. Michael Sheffery is a recipient of a Miriam and Benedict Wolf Fellowship.

REFERENCES

Allan, M.; Grindlay, G. J.; Stefani, L.; and Paul, J. (1982). Epsilon globin gene transcripts originating upstream of the mRNA cap site in mK562 cells and normal human embryos. *Nucl. Acids Res.* **10,** 5133–5147.

Allan, M.; Lanyon, W. G.; and Paul, J. (1983). Multiple origins of transcription in the 4.5kb upstream of the Σ-globin gene. *Cell* **35,** 187–197.

Andersson, K.; Mahr, R.; Bjorkroth, B.; and Daneholt, B. (1982). Rapid reformation of the thick chromosome fiber upon completion of RNA synthesis at the Balbiani ring genes in Chironomus tentans. *Chromosoma* **87,** 33–48.

Axel, R. (1975). Cleavage of DNA in nuclei and chromatin with staphylococcal nuclease. *Biochemistry* **14,** 2921-2925.

Baer, B. W.; and Rhodes, D. (1983). Eucaryotic RNA polymerase II binds to nucleosome cores from transcribed genes. *Nature* **301,** 482-488.

Balcarek, J. M.; and McMorris, F. A. (1983). DNase I hypersensitive sites of globin genes of uninduced Friend Erythroleukemia Cells and changes during induction with dimethyl sulfoxide. *J. Biol. Chem.* **258,** 10622-10628.

Ball, D. J.; Gross, D. S.; and Garrard, W. T. (1983). 5-Methylcytosine is localized in nucleosomes that contain histone H1. *Proc. Natl. Acad. Sci. (U.S.A.)* **80,** 5490-5494.

Banerji, J.; Rusconi, S.; and Schaffner, W. (1983). Expression of a B-globin gene is enhanced by remote SV40 DNA sequences. *Cell* **27,** 299-308.

Banerji, J.; Olson, L.; and Schaffner, W. (1983). A lymphocyte-specific cellular enhancer is located downstream of the joining region in immunoglobulin heavy chain genes. *Cell* **33,** 729-740.

Behe, M.; and Felsenfeld, G. (1981). Effects of methylation on a synthetic polynucleotide: The B-Z transition in poly (dG-m^{5}dC) poly (dG-m^{5}dC). *Proc. Natl. Acad. Sci. (U.S.A.)* **78,** 1619-1623.

Bellard, M.; Gannon, F.; and Chambon, P. (1977). Nucleosome structure III: The structive and transcriptional activity of the chromatin containing the ovalbumin and globin genes in chick oviduct nuclei. *Cold Spring Harbor Symp. Quant. Biol.* **42,** 779-791.

Bellard, M.; Kuo, M. T.; Dretzen, G.; and Chambon, P. (1980). Differential nuclease sensitivity of the ovalbumin and B-globin chromatin regions in erythrocytes and oviduct cells of laying hens. *Nucl. Acids Res.* **8,** 2737-2750.

Bellard, M.; Dretzen, G.; Bellard, F.; Oudet, P.; and Chambon, P. (1982). Disruption of the typical chromatin-structure in a 2500 base-pair region at the 5' end of the actively transcribed ovalbumin gene. *EMBO J.* **1,** 223-230.

Bergman, L. W.; and Kramer, R. A. (1983). Modulation of chromatin structure associated with derepression of the acid phosphatase gene of Saccharromyces cerevisiae. *J. Biol. Chem.* **258,** 7223-7227.

Birchmeier, C.; Grosschedl, R.; and Birnstiel, M. L. (1982). Generation of authentic 3' termini of an H2A mRNA *in vivo* is dependent on a short inverted DNA repeat and on spacer sequences. *Cell* **28,** 739-745.

Bird, A. P. (1984). DNA methylation—how important in gene control? *Nature* **307,** 503-504.

Bird, A. P.; and Southern, E. M. (1978). Use of restriction enzymes to study eukaryotic DNA methylation. *J. Mol. Biol.* **118,** 27-47.

Bitter, G. A.; and Roeder, R. G. (1978). Transcription of viral genes by RNA polymerase II in nuclei isolated from adenovirus 2 transformed cells. *Biochemistry* **17,** 2198-2205.

Bloom, K. S.; and Anderson, J. N. (1978). Fractionation of hen oviduct chromatin into transcriptionally active and inactive regions after selective micrococcal nuclease digestion. *Cell* **15,** 141-150.

Bloom, K. S.; and Anderson, J. N. (1979). Conformation of ovalbumin and globin genes in chromatin during differential gene expression. *J. Biol. Chem.* **254,** 10532-10539.

Bonven, B.; and Westergaard, O. (1982). DNAase I hypersensitive regions correlate with a site specific endogenous nuclease activity in the r-chromatin of Tetrahymena. *Nucl. Acids Res.* **10,** 7593-7608.

Borchsenius, S.; Bonven, B.; Leer, J. C.; and Westergaard, O. (1981). Nuclease sensitive regions on the extrachromosomal r-chromatin from Tetrahymens pyriformis. *Eur. J. Biochem.* **117,** 245-250.

Bryan, P. N.; Hofsetter, H.; and Birnstiel, M. L. (1981). Nucleosome arrangement on tRNA genes of *Xenopus laevis*. *Cell* **27,** 459-466.

Bryan, P. N.; Olah, J.; and Birnstiel, M. L. (1983). Major changes in the 5' and 3' chromatin structure of sea urchin histone genes accompany their activation and inactivation in development. *Cell* **33**, 843-848.

Busslinger, M.; Hurst, J.; and Flavell, R. A. (1983). DNA methylation and the regulation of globin gene expression. *Cell* **34**, 197-206.

Burch, J. B. E. (1984). Identification and sequence analysis of the 5' end of the major chicken vitellogenin gene. *Nucl. Acids Res.* **12**, 1117-1135.

Burch, J. B. E.; and Weintraub, H. (1983). Temporal order of chromatin structural changes associated with activation of the major chicken vitellogenin gene. *Cell* **33**, 65-76.

Carlson, D. P.; and Ross, J. (1983). Human B-globin promoter and coding sequences transcribed by RNA polymerase III. *Cell* **34**, 857-864. Cartwright, I. L.; and Elgin, S. C. R. (1982). Analysis of chromatin structure and DNA sequence organization: use of the 1,0 phenanthroline-cuprous complex. *Nucl. Acids Res.* **10**, 5835-5852.

Cartwright, I. L.; Hertzberg, R. P.; Dervan, P. B.; and Elgin, S. C. R. (1983). Cleavage of chromatin with methidiumproply-EDTA-iron(II). *Proc. Natl. Acad. Sci. (U.S.A.)* **80**, 3213-3217.

Cate, R. L.; Chick, W.; and Gilbert, W. (1983). Comparison of the methylation patterns of the two rat insulin genes. *J. Biol. Chem.* **258**, 6645-6652.

Chao, M. V.; Mellon, P.; Charnay, P.; Maniatis, T.; and Axel, R. (1983). The regulated expression of B-globin genes introduced into mouse erythroleukemia cells. *Cell* **32**, 483-493.

Charnay, P.; and Maniatis, T. (1983). Transcriptional regulation of globin gene expression in the human erythroid cell line K562. *Science* **220**, 1281-1283.

Charnay, P.; Treisman, R.; Mellon, P.; Chao, M.; Axel, R.; and Maniatis, T. (1984). Differences in human α and B-globin gene expression in mouse erythroleukemia cells: the role of intragenic sequences. *Cell* **38**, 251-263.

Christman, J. K.; Weich, N.; Schoenbrun, B.; Schneiderman, N.; and Acs, G. (1980). Hypomethylation of DNA during differentiation of Friend erythroleukemia cells. *J. Cell Biol.* **86**, 366-370.

Chung, S.-Y.; Folsom, V.; and Wooley, J. (1983). DNase I hypersensitive sites in the chromatin of immunoglobulin kappa light chain genes. *Proc. Natl. Acad. Sci. (U.S.A.)* **80**, 2427-2431.

Church, G. M.; and Gilbert, W. (1984). Genome sequencing. *Proc. Natl. Acad. Sci. (U.S.A.)* **81**, 1991-1995.

Cockerill, P. N.; and Goodwin, G. H. (1983). Demonstration of an S1-nuclease sensitive site near the human B-globin gene, and its protection by HMG 1 and 2. *Biochem. and Biophys. Res. Commun.* **112**, 547-554.

Cohen, R. B.; and Sheffery, M. (1985). Nucleosome disruption precedes transcription and is largely limited to the transcribed domain of globin genes in murine erythroleukemia cells. *J. Mol. Biol.* **182**, 109-129.

Compere, S. J.; and Palmiter, R. D. (1981). DNA methylation controls the inducibility of mouse metallothionein-I gene in lymphoid cells. *Cell* **25**, 233-240.

Creusot, F.; Acs, G.; Christman, J. K. (1982). Inhibition of DNA methyltransferase and induction of Friend erythroleukemia cell differentiation by 5-aza cytidine and 5-AZA-2'-deoxycytidine. *J. Biol. Chem.* **257**, 2041-2048.

Darnell, J. E., Jr. (1982). Variety in the level of gene control in eukaryotic cells. *Nature* (London) **297**, 365-371.

Davis, A. H.; Reudelhuber, T. L.; and Garrard, W. T. (1983). Variegated chromatin structures of mouse ribosomal RNA genes. *J. Mol. Biol.* **167**, 133-155.

Derman, E. (1981). Isolation of a cDNA clone for mouse urinary proteins: Age- and sex-related expression of mouse urinary protein genes is transcriptionally controlled. *Proc. Natl. Acad., Sci. (U.S.A.)* **78**, 5425-5429.

Derman, E.; Krauter, K.; Walling, L.; Weinberger, C.; Ray, M.; and Darnell, J. E., Jr. (1981). Transcriptional control in the production of liver-specific mRNAs. *Cell* **23**, 731-739.

Dierks, P.; van Ooyen, A.; Cochran, M. D.; Dobkin, C.; Reiser, J.; and Weissmann, C. (1983). Three regions upstream from the cap site are required for efficient and accurate transcription of the rabbit *B*-globin gene in mouse 3T3 cells. *Cell* **32**, 695-706.

Dingwall, C.; Lomonossoff, G. P.; and Lasky, R. A. (1981). High sequences specificity of micrococcal nuclease. *Nucl. Acids Res.* **9**, 2659-2673.

Doerfler, W. (1983). DNA methylation and gene activity. *Annu. Rev. Biochem.* **52**, 93-124.

Efstradiadis, A.; Posakony, J. W.; Maniatis, T.; Lawn, R. M.; O'Connell, C.; Spritz, R. A.; DeRiel, J. K.; Forget, B. G.; Weissmann, S. M.; Slightom, J. L.; Blechl, A. E.; Smithies, O.; Baralle, F. E.; Shoulders, C. C.; and Proudfoot, N. J. (1980). The structure and evolution of the human *B*-globin gene family. *Cell* **21**, 563-668.

Elgin, S. C. R. (1981). DNAase I-hypersensitive sites of chromatin. *Cell* **27**, 413-415.

Elgin, S. C. R. (1984). Anatomy of hypersensitive sites. *Nature* **309**, 213-214.

Emerson, B. M.; and Felsenfeld, G. (1984). Specific factor conferring nuclease hypersensitivity at the 5' end of the chicken adult *B*-globin gene. *Proc. Natl. Acad. Sci. (U.S.A.)* **81**, 95-99.

Finer, M. H.; Fodor, E. J. B.; Boedtker, H.; and Doty, P. (1984). Endonuclease S1-sensitive site in chicken pro-α2(I) collagen 5' flanking gene region. *Proc. Natl. Acad. Sci. (U.S.A.)* **81**, 1659-1663.

Folger, K.; Anderson, J. N.; Hayward, M. A.; and Shapiro, D. J. (1983). Nuclease sensitivity and DNA methylation in estrogen regulation of *Xenopus laevis* vitellogenin gene expression. *J. Biol. Chem.* **258**, 8908-8914.

Fradin, A.; Manley, J. L.; and Prives, C. L. (1982). Methylation of simian virus 40 Hpa II-site affects late but not early viral gene expression. *Proc. Natl. Acad. Sci. (U.S.A.)* **79**, 5142-5246.

Fritton, H. P.; Sippel, A. E.; and Igo-Kemenes, T. (1983). Nuclease-hypersensitive sites in the chromatin domain of the chicken lysozyme gene. *Nucl. Acids Res.* **11**, 3467-3485.

Fromm, M.; and Berg, P. (1983). Simian virus 40 early- and late-region promoter functions are enhanced by the 72-base pair repeat inserted at distant locations and inverted orientations. *Mol. Cell. Biol.* **3**, 991-999.

Gall, J. G.; and Callan, H. G. (1962). H^3 Uridine incorporation in lampbrush chromosomes. *Proc. Natl. Acad. Sci. (U.S.A.)* **48**, 562-570.

Garel, A.; and Axel, R. (1976). Selective digestion of transcriptionally active ovalbumin genes from oviduct nuclei. *Proc. Natl. Acad. Sci. (U.S.A.)* **73**, 3966-3970.

Gariglio, P.; Bellard, M.; and Chambon, P. (1981). Clustering of RNA polymerase B molecules in the 5' moiety of the adult *B*-globin gene of hen erythrocytes. *Nucl. Acids Res.* **9**, 2589-2598.

Gazit, B.; Panet, A.; and Cedar, H. (1980). Reconstitution of a deoxyribonuclease I-sensitive structure on active genes. *Proc. Natl. Acad. Sci. (U.S.A.)* **77**, 1787-1790.

Germond, J. E.; Hirt, B.; Oudet, P.; Gross-Bellard, M.; and Chambon, P. (1975). Folding of the DNA double helix in chromatin-like structures from Simian virus 40. *Proc. Natl. Acad. Sci. (U.S.A.)* **72**, 1843-1847.

Gillies, S. D.; Morrison, S. L.; Oi, V. T.; and Tonegawa, S. (1983). A tissue-specific transcription enhancer element is located in the major intron of a rearranged immunoglobulin heavy chain gene. *Cell* **33**, 717-728.

Glikin, G. C.; Ruberti, I.; and Worcel, A. (1984). Chromatin assembly in Xenopus Oocytes: *in vitro* studies. *Cell* **37**, 33-41.

Goodwin, G. H.; Mathew, C. G. P.; Wright, C. A. W.; Venkov, C.; and Johns, E. W. (1979). Analysis of the high mobility group proteins associated with salt-soluble nucleosomes. *Nucl. Acids Res.* **7**, 1815-1835.

Goodwin, G. H.; Wright, C. A.; and Johns, E. W. (1981). The characterization of 1SF monomer nucleosomes from hen oviduct and the partial characterization of a third HMG 14/17-like protein in such nucleosomes. *Nucl. Acids Res.* **9**, 2761-2775.

Gottesfeld, J. M.; and Butler, P. J. G. (1977). Structure of transcriptionally active chromatin subunits. *Nucl. Acids Res.* **4,** 3155-3173.

Grainger, R. M.; Hazard-Leonards, R. M.; Samaha, F.; Hougan, L. M.; Lesk, M. R.; and Thomsen, G. H. (1983). Is hypomethylation linked to activation of λ-crystallin genes during development? *Nature* (London) **306,** 88-91.

Green, M. R.; Treisman, R.; and Maniatis, T. (1983). Transcriptional activation of cloned human *B*-globin genes by viral immediate-early gene products. *Cell* **35,** 137-148.

Gregory, S. P.; and Butterworth, P. H. W. (1983). A comparison of the promoter strengths of two eukaryotic genes *in vitro* reveals a region of DNA that can influence the rate of transcription in *cis* over long distances. *Nucl. Acids Res.* **11,** 5317-5326.

Grosveld, G. C.; Shewmaker, C. K.; Jat, P.; and Flavell, R. A. (1981). Localization DNA sequences necessary for transcription of the rabbit *B*-globin gene *in vitro*. *Cell* **25,** 215-226.

Grosveld, G. C.; deBoer, E.; Shewmaker, C. K.; and Flavell, R. A. (1982). DNA sequences necessary for transcription of the rabbit *B*-globin gene in vivo. *Nature* (London) **295,** 120-126.

Groudine, M.; and Weintraub, H. (1981). Activation of globin genes during chicken development. *Cell* **24,** 393-401.

Groudine, M.; and Weintraub, H. (1982). Propagation of globin DNAase I hypersensitive sites in absence of factors required for induction: A possible mechanism for determination. *Cell* **30,** 131-139.

Groudine, M.; Peretz, M.; and Weintraub, H. (1981a). Transcriptional regulation of hemoglobin switching in chicken embryos. *Mol. Cell. Biol.* **1,** 281-288.

Groudine, M.; Eisenman, R.; and Weintraub, H. (1981b). Chromatin structure of endogenous retro-viral genes and activation by an inhibitor of DNA methylation. *Nature* **292,** 311-317.

Groudine, M.; Kohwi-Shigematsu, T.; Gelinas, R.; Stamatoyannopoulos, G.; and Papayannopoulou, T. (1983). Human fetal to adult hemoglobin switching: changes in chromatin structure of the *B*-globin gene locus. *Proc. Natl. Acad. Sci. (U.S.A.)* **80,** 7551-7555.

Harbers, K.; Schieke, A.; Stuhlmann, H.; Jahner, D.; and Jaenisch, (1981). DNA methylation and gene expression: endogenous retroviral genome becomes infectious after molecular cloning. *Proc. Natl. Acad. Sci. (U.S.A.)* **78,** 7609-7613.

Harland, R. M.; Weintraub, H.; and McKnight, S. L. (1983). Transcription of DNA injected into Xenopus oocytes is influenced by template topology. *Nature* **302,** 38-43.

Herbomel, P.; Saragosti, S.; Blangy, D.; and Yaniv, M. (1981). Fine structure of the origin-proximal DNAase I hypersensitive region in mild-type and EC mutant polyoma. *Cell* **25,** 651-658.

Hereford, L.; Bromley, S.; and Osley, M. A. (1982). Periodic transcription of yeast histone genes. *Cell* **30,** 305-310.

Hofer, E.; and Darnell, J. E., Jr. (1981). The primary transcription unit of the mouse *B*-major globin gene. *Cell* **23,** 585-593.

Hofer, E.; Hofer-Warbinek, R.; and Darnell, J. E., Jr. (1982). Globin RNA transcription: A possible termination site and demonstration of transcriptional control correlated with altered chromatin structure. *Cell* **29,** 887-893.

Holliday, R.; and Pugh, J. E. (1975). DNA modification mechanisms and gene activity during development. *Science* **187,** 226-232.

Horz, W.; and Altenburger, W. (1981). Sequence specific cleavage of DNA by micrococcal nuclease. *Nucl. Acids. Res.* **9,** 2643-2658.

Humphries, R. K.; Ley, T.; Turner, P.; Moulton, A. D.; and Nienhuis, A. W. (1982). Differences in human α-, *B*-, and λ-globin gene expression in monkey kidney cells. *Cell* **30,** 173-183.

Igo-Kemenes, T.; Horz, W.; and Zachau, H. G. (1982). Chromatin. *Annu. Rev. Biochem.* **51,** 89-121.

Jahner, D.; Stuhlmann, H.; Stewart, C. L.; Harbers, K.; Lohler, J.; Simon, I.; and Jaenisch, R. (1982). De novo methylation and expression of retroviral genomes during mouse embryogenesis. *Nature* **298,** 623-628.

Jakobovits, E. B.; Bratosin, S.; and Aloni, Y. (1980). A nucleosome-free region in SV40 minichromosomes. *Nature* **285**, 263-265.

Javaherian, K.; and Liu, L. F. (1983). Association of eukaryotic DNA Topoisomerase I with nucleosomes and chromosomal proteins. *Nucl. Acids Res.* **11**, 461-472.

Javaherian, K.; and Fasman, G. D. (1984). Nick tranlsation of HeLa cell nuclei as a probe for locating DNase I-sensitive nuclesomes. *J. Biol. Chem.* **259**, 3343-3349.

Jessee, B.; Gargiulo, G.; Razvi, F.; and Worcel, A. (1982). Analogous cleavage of DNA by micrococcal nuclease and a 1,10-phenanthroline-cuprous complex. *Nucl. Acids Res.* **10**, 5823-5834.

Johnson, E. M.; Allfrey, V. G.; Bradbury, E. M.; and Mathews, H. R. (1978). Altered nucleosome structure containing DNA sequences complementary to 19S and 26S ribosomal RNA in Physarum. *Proc. Natl. Acad. Sci. (U.S.A.)* **75**, 1116-1120.

Johnson, E. M.; Campbell, G. R.; and Allfrey, V. G. (1979). Different nucleosome structures on transcribing and nontranscribing ribosomal gene sequences. *Science* **206**, 1192-1193.

Jones, P. A.; and Taylor, S. M. (1980). Cellular differentiation, cytidine analogs, and DNA methylation. *Cell* **20**, 85-93.

Jones, P. A.; and Taylor, S. M. (1981). Hemimethylated duplex DNAs prepared from 5-aza-cytidine-treated cells. *Nucl. Acids Res.* **9**, 2933-2947.

Jongstra, J.; Reudelhuber, T. L.; Oudet, P.; Benoist, C.; Chae, C.-B.; Jeltsch, J.-M.; Mathis, D. J.; and Chambon, P. (1984). Induction of altered chromatin structures by simian virus 40 enhancer and promoter elements. *Nature* (London) **307**, 708-714.

Karpov, V. L.; Preobazhenskaya, O. V.; and Mirzabekov, A. D. (1984). Chromatin structure of hsp 70 genes, activated by heat shock: selective removal of histones from the coding region and their absence from the 5' region. *Cell* **36**, 423-431.

Kaye, J. S.; Bellard, M.; Dretzen, G.; Bellard, F.; and Chambon, P. (1984). A close association between sites of DNase I hypersensitivity and sites of enhanced cleavage by micrococcal nuclease in the 5'-flanking region of the actively transcribed ovalbumin gene. *EMBO J.* **3**, 1137-1144.

Keene, M. A.; and Elgin, S. C. R. (1981). Micrococcal nuclease as a probe of DNA sequence organization and chromatin structure. *Cell* **27**, 57-64.

Keene, M. A.; Corces, V.; Lowenhaupt, K.; and Elgin, S. C. R. (1981). DNase hypersensitive sites in Drosophila chromatin occur at the 5' ends of regions of transcription. *Proc. Natl. Acad. Sci. (U.S.A.)* **78**, 143-146.

Khoury, G.; and Gruss, P. (1983). Enhancer elements. *Cell* **33**, 313-314.

Kimura, T.; Mills, F. C.; Allan, J.; and Gould, H. (1983). Selective unfolding of erythroid chromatin in the region of the active B-globin gene. *Nature* **306**, 709-712.

Kohwi-Shigematsu, T.; Gelinas, R.; and Weintraub, H. (1983). Detection of an altered DNA conformation at specific sites in chromatin and supercoiled DNA. *Proc. Natl. Acad. Sci. (U.S.A.)* **80**, 4389-4393.

Koostra, A. (1982). Isolation of high mobility group-containing mononucleosomes from avian erythrocyte nuclei and their sensitivity to DNase I. *J. Biol. Chem.* **257**, 13088-13094.

Kruczek, I.; and Doerfler, W. (1983). Expression of the chloramphenicol acetyltransferase gene in mammalian cells under the control of adenovirus type 12 promoters: Effect of promoter methylation on gene expression. *Proc. Natl. Acad. Sci. (U.S.A.)* **80**, 7586-7590.

Kuo, M. T.; Mandel, J. S.; and Chambon, P. (1979). DNA methylation: correlation with DNase I sensitivity of chicken ovalbumin and conalbumin chromatin. *Nucl. Acids Res.* **7**, 2105-2113.

Labhart, P.; and Koller, T. (1982). Structure of the active nucleolar chromatin of Xenopus laevis oocytes. *Cell* **28**, 279-292.

Lacy, E.; and Axel, R. (1975). Analysis of DNA of isolated chromatin subunits. *Proc. Natl. Acad. Sci. (U.S.A.)* **72**, 3978-3982.

Landes, G. M.; Villeponteau, B.; Pribyl, T. M.; and Martinson, H. G. (1982). Hemoglobin switching in chickens: is the switch initiated post-transcriptionally? *J. Biol. Chem.* **257,** 11008-11014.

Larsen, A.; and Weintraub, H. (1982). An altered DNA conformation detected by S1 nuclease occurs at specific regions in active chick globin chromatin. *Cell* **29,** 609-622.

Lawson, G. M.; Tsai, M.-J.; and O'Malley, B. W. (1980). Deoxyribonuclease I sensitivity of the nontranscribed sequences flanking the 5' and 3' ends of the ovomucoid gene and the ovalbumin and its related X and Y genes in hen oviduct nuclei. *Biochemistry* **19,** 4403-4411.

Lawson, G. M.; Knoll, B. J.; March, C. J.; Woo, S. L. C.; Tsai, M.-J.; and O'Malley, B. W. (1982). Definition of 5' and 3' structural boundaries of the chromatin domain containing the ovalbumin multigene family. *J. Biol. Chem.* **257,** 1501-1507.

Letvin, N. L.; Linch, D. C.; Beardsley, G. P.; McIntyre, K. W.; and Nathan, D. G. (1984). Augmentation of fetal-hemoglobin production in anemic patients by hydroxyurea. *N. Engl. J. Med.* **310,** 869-873.

Levinger, L.; and Varshavsky, A. (1982). Selective arrangement of ubiquitinated and D1 protein-containing nucleosomes within the Drosophila genome. *Cell* **28,** 375-385.

Levinger, L.; Barsoum, J.; and Varshavsky, A. (1981). Two-dimensional hyridization mapping of nucleosomes: comparison of DNA and protein patterns. *J. Mol. Biol.* **146,** 287-304.

Levitt, A.; Axel, R.; and Cedar, H. (1979). Nick translation of active genes on intact nuclei. *Dev. Biol.* **69,** 496-505.

Levy, A.; and Noll, M. (1981). Chromatin fine structure of active and repressed genes. *Nature* **289,** 198-203.

Levy, B.; Connor, W.; and Dixon, G. (1979). A subset of trout testis nucleosomes enriched in transcribed DNA sequences contains high mobility group proteins as major structural components. *J. Biol. Chem.* **254,** 609-620.

Ley, T. J.; DeSimone, J., Anagnon, N.P.; Keller, G. H.; Humphries, R. K.; Turner, P. H.; Young, N. S.; Heller, P.; and Nienhuis, A. W. (1982). 5-Azacytidine selectively increases γ-globin synthesis in a patient with $B+$ thalassemia. *N. Engl. J. Med.* **307,** 1469-1475.

Lieberman, M. W.; Beach, L. R.; and Palmiter, R. D. (1983). Ultraviolet radiation-induced metalothionein-I gene activation is associated with extensive DNA demethylation. *Cell* **35,** 207-214.

Lowenhaupt, K.; Trent, C.; and Lingrel, J. B. (1978). Mechanisms for accumulation of globin mRNA during dimethyl sulfoxide induction of murine erythroleukaemia cells: synthesis of precursors and mature mRNA. *Dev. Biol.* **63,** 441-454.

Luchnik, A.; Bakayev, V. V.; Zbarsky, I. B.; and Georgiev, G. P.; (1982). Elastic torsional strain in DNA within a fraction of SV40 minichromosomes: relation to transcriptionally active fraction. *EMBO J.* **1,** 1353-1358.

Mace, H. A. F.; Pelham, H. R. B.; and Travers, A. A. (1983). Association of an S1 nuclease sensitive structure with short direct repeats 5' of Drosophila heat shock genes. *Nature* **304,** 555-557.

McGhee, J. D.; and Ginder, G. D. (1979). Specific DNA methylation sites in the vicinity of the chicken *B*-globin gene. *Nature* **280,** 419-420.

McGhee, J. D.; Wood, W. I.; Dolan, M.; Engel, J. D.; and Felsenfeld, G. (1981). A 200 base pair region at the 5' end of the chicken adult *B*-globin gene is accessible to nuclease digestion. *Cell* **27,** 45-55.

McGinnis, W.; Shermoen, A. W.; Heemskerk, J.; and Beckendorf, S. K. (1983*a*). DNA sequence changes in an upstream DNase I-hypersensitive region are correlated with reduced gene expression. *Proc. Natl. Acad. Sci. (U.S.A.)* **80,** 1063-1067.

McGinnis, W.; Shermoen, A. W.; and Beckendorf, S. K. (1983*b*). A transposable element inserted just 5' to a Drosophila glue protein gene alters gene expression and chromatin structure. *Cell* **34,** 75-84.

McKeon, C.; Ohkubo, H.; Pastan, I.; and deCrombrugghe, B. (1982). Unusual methylation pattern of the α2(I) collagen gene. *Cell* **29**, 203-210.

McKeon, C.; Pastan, I.; and deCrombrugghe, B. (1984a). DNase I sensitivity of the α2(I) collagen gene: correlation with its expression but not with its methylation pattern. *Nucl. Acids Res.* **12**, 3491-3502.

McKeon, C.; Schmidt, A.; and deCrombrugghe, B. (1984b). A sequence conserved in both the chicken and mouse α2(I) collagen promoter contains sites sensitive to S1 nuclase. *J. Biol. Chem.* **259**, 6636-6640.

McKnight, G. S.; and Palmiter, R. D. (1979). Transcriptional regulation of the ovalbumin and conalbumin genes by steroid hormones in chick oviduct. *J. Biol. Chem.* **254**, 9050-9058.

McKnight, S. L.; Martin, K. A.; Beyer, A. L.; and Miller, O. L., Jr. (1979). Visualization of functionally active chromatin. In *The Cell Nucleus* (H. Busch, ed.), vol. 7, pp. 97-122, Academic Press, New York.

Macleod, D.; and Bird, A. P. (1982). DNAase I sensitivity and methylation of active versus inactive rRNA genes in Xenopus species hybrids. *Cell* **29**, 211-218.

Macleod, D.; and Bird,. A. P. (1983). Transcription in oocytes of highly methylated rDNA from Xenopus laevis sperm. *Nature* **306**, 200-203.

Marks, P. A.; and Rifkind, R. A. (1978). Erythroleukemic differentiation. *Annu. Rev. Biochem.* **47**, 419-448.

Mather, E. L.; and Perry, R. (1983). Methylation status and DNAase I sensitivity of immunoglobulin genes: changes associated with rearrangement. *Proc. Natl. Acad. Sci. (U.S.A.)* **80**, 4689-4693.

Mathis, D. J.; and Gorovsky, M. A. (1977). Structure of rDNA-containing chromatin of Tetrahymena pyriformis analyzed by nuclease digestion. *Cold Spring Harbor Symp. Quant. Biol.* **42**, 773-778.

Mellon, P.; Parker, V.; Gluzman, Y.; and Maniatis, T. (1981). Identification of DNA sequences required for transcription of the human α1-globin gene in a new SV40 host-vector system. *Cell* **27**, 279-288.

Miller, D. M.; Turner, P.; Nienhuis, A. W.; Axelrod, D. E.; and Gopalakrishnan, T. V. (1978). Active conformation of the globin genes in uninduced and induced mouse erythroleukemia cells. *Cell* **14**, 511-524.

Mills, F. C.; Fisher, L. M.; Kuroda, R.; Ford, A. M.; and Gould, H. J. (1983). DNase I hypersensitive sites in the chromatin of human mu immunoglobulin heavy-chain genes. *Nature* **306**, 809-812.

Nasmyth, K. A. (1982). The regulation of yeast mating-type chromatin structure by SIR: an action at a distance affecting both transcription and transposition. *Cell* **30**, 567-578.

Nedospasov, S. A.; and Georgiev, G. P. (1980). Nonrandom cleavage of SV40 DNA in compact minichromosomes and free in solution by M. nuclease. *Biochem. Biophys. Res. Commun.* **92**, 532-539.

Ness, P. J.; Labhart, P.; Banz, E.; Koller, T.; and Parish, R. W. (1983). Chromatin structure along the ribosomal DNA of Dictyostelium: Regional differences and changes accompanying cell differentiation. *J. Mol. Biol.* **166**, 361-381.

Nickol, J. M.; Behe, M.; and Felsenfeld, G. (1982). Effect of the B-Z transition in poly (dG-m^{5}dC) on nucleosome formation. *Proc. Natl. Acad. Sci. (U.S.A.)* **79**, 1771-1775.

Nickol, J. M.; and Felsenfeld, G. (1983). DNA conformation at the 5' end of the chicken adult β-globin gene. *Cell* **35**, 467-477.

Nicolas, R. H.; Wright, C. A.; Cockerill, P. N.; Wyke, J. A.; and Goodwin, G. H. (1983). The nuclease sensitivity of active genes. *Nucl. Acid Res.* **11**, 753-772.

Ott, M. O.; Sperling, L.; Cassio, D.; Levilliers, J.; Salat-Trepat, J.; and Weiss, M. C. (1982). Undermethylation at the 5' end of the albumin gene is necessary but not sufficient for albumin production by rat hepatoma cells in culture. *Cell* **30**, 825-833.

Ott, M. O.; Sperling, L.; and Weiss, M. C. (1984). Albumin extraction without mehylation of its gene. *Proc. Natl. Acad. Sci. (U.S.A.)* **81**, 1738-1741.
Palen, T. E.; and Cech, T. R. (1984). Chromatin structure at the replication origins and transcription-initiation regions of the ribosomal RNA genes of tetrahymena. *Cell* **36**, 933-942.
Parslow, T. G.; and Granner, D. K. (1983a). Chromatin changes accompanying immunoglobulin kappa gene activation: a potential control region without the gene. *Nature* **299**, 449-451.
Parslow, T. G.; and Granner, D. K. (1983b). Structure of a nuclease-sensitive region inside the immunoglobulin kappa chain gene: evidence for a role in gene regulation. *Nucl. Acids Res.* **11**, 4775-4792.
Paulson, J. R.; and Laemmli, U. K. (1977). The structure of histone depleted metaphase chromosomes. *Cell* **12**, 817-828.
Pelham, H. R. B. (1982). A regulatory upstream promoter element in the Drosophila Hsp 70 heat-shock gene. *Cell* **30**, 517-528.
Pfeiffer, W.; and Zachau, H. G. (1980). Accessibility of expressed and nonexpressed genes to a restriction endonuclease. *Nucl. Acids Res.* **8**, 4621-4638.
Picard, D.; and Schaffner, W. (1984). A lymphocyte-specific enhancer in the mouse immunoglobulin kappa gene. *Nature* **307**, 80-82.
Prior, C. P.; Cantor, C. R.; Johnson, E. M.; Littau, V. C.; and Allfrey, V. G. (1983). Reversible changes in nucleosome structure and histone H3 accessibility in transcriptionally active and inactive states of rDNA chromatin. *Cell* **34**, 1033-1042.
Profous-Juchelka, H. R.; Reuben, R. C.; Marks, P. A.; and Rifkind, R. A. (1983). Transcriptional and post-transcriptional regulation of globin gene accumulation in murine erythroleukemia cells. *Mol. Cell. Biol.* **3**, 229-232.
Pruitt, S. C.; and Reeder, R. H. (1984). Effect of topological constraint on transcription of ribosomal DNA in Xenopus oocytes. *J. Mol. Biol.* **174**, 121-139.
Queen, C.; and Baltimore, D. (1983). Immunoglobulin gene transcription is activated by downstream sequence elements. *Cell* **33**, 741-748.
Razin, A.; and Riggs, A. D. (1980). DNA methylation and gene function. *Science* **210**, 604-610.
Reeves, R. (1978). Nucleosome structure of Xenopus oocyte amplified ribosomal genes. *Biochemistry* **17**, 4908-4916.
Reeves, R.; and Jones, A. (1976). Genomic transcriptional activity and the structure of chromatin. *Nature* **260**, 495-500.
Reudelhuber, T. L.; Ball, D. J.; Davis, A. H.; and Garrard, W. T. (1982). Transferring DNA from electrophoretically resolved nucleosomes to diazobenzyloxmethyl cellulose: properties of nucleosomes along mouse satellite DNA. *Nucl. Acids Res.* **10**, 1311-1325.
Rose, S. M.; and Garrard, W. T. (1984). Differentiation-dependent chromatin alterations precede and accompany transcription of immunoglobulin light chain genes. *J. Biol. Chem.* **259**, 8534-8544.
Ryoji, M.; and Worcel, A. (1984). Chromatin assembly in Xenopus oocytes: in vivo studies. *Cell* **37**, 21-32.
Salditt-Georgieff, M.; and Darnell, J. E., Jr. (1983). A precise termination site in the mouse B-major globin transcription unit. *Proc. Natl. Acad. Sci. (U.S.A.)* **80**, 4694-4698.
Salditt-Georgieff, M.; Sheffery, M.; Krauter, K.; Darnell, J. E., Jr.; Rifkind, R. A.; and Marks, P. A. (1984). Induced transcription of the mouse B-globin transcription unit in erythroleukameia cells: time course of induction and changes in chromatin structure. *J. Mol. Biol.* **172**, 437-450.
Samal, B.; Worcel, A.; Louis, C.; and Schedl, P. (1981). Chromatin structure of the histone genes of D. melanogaster. *Cell* **23**, 401-409.
Sandeen, G.; Wood, W. I.; and Felsenfeld, G. (1980). The interaction of high mobility proteins HMGs 14 and 17 with nucleosomes. *Nucl. Acids Res.* **8**, 3757-3778.

Saragosti, S.; Moyne, G.; and Yaniv, M. (1980). Absence of nucleosomes in a fraction of SV40 chromatin between the origin of replication and the region coding for the late leader RNA. *Cell* **20**, 65-73.

Saragosti, S.; Cereghini, S.; and Yaniv, M. (1982). Fine structure of the regulatory region of Simian virus 40 minichromosomes revealed by DNAase I digestion. *J. Mol. Biol.* **160**, 133-146.

Schibler, U.; Hagenbuchle, O.; Wellauer, P. K.; and Pittet, A. C. (1983). Two promoters of different strengths control the transcription of the mouse alpha-amylase gene Amy -1^a in the parotid gland and the liver. *Cell* **33**, 501-508.

Scholer, H. R.; and Gruss, P. (1984). Specific interaction between enhancer-containing molecules and cellular components. *Cell* **35**, 837-848.

Schubach, W.; and Groudine, M. (1984). Alteration of c-myc chromatin structure by avian leukososis virus integration. *Nature* **307**, 702-708.

Scott, W. A.; and Wigmore, D. J. (1978). Sites in simian virus 40 chromatin which are preferentially cleaved by endonucleases. *Cell* **15**, 1511-1518.

Senear, A. W.; and Palmiter, R. D. (1981). Multiple structural features are responsible for the nuclease sensitivity of the active ovalbumin gene. *J. Biol. Chem.* **256**, 1191-1198.

Sheffery, M.; Rifkind, R. A.; and Marks, P. A. (1982). Murine erythroleukemia cell differentiation: DNase I hypersensitivity and DNA methylation near the globin genes. *Proc. Natl. Acad. Sci., (U.S.A.)* **79**, 1180-1184.

Sheffery, M.; Rifkind, R. A.; and Marks, P. A. (1983). Hexamethylenebisacetamide-resistant murine erytholeukemia cells have altered patterns of inducer-mediated chromatin changes. *Proc. Natl. Acad. Sci. (U.S.A.)* **80**, 3349-3353.

Sheffery, M.; Marks, P. A.; and Rifkind, R. A. (1984). Gene expression in murine erythroleukemia cells: transcriptional control and chromatin structure of the a_1-globin gene. *J. Mol. Biol.* **172**, 417-436.

Shen, C.-K. J. (1983). Superhelicity induces hypersensitivity of a human polypyrimidien polypurine DNA sequence in the human a_2-a_1 globin intergenic region to S1 nuclease digestion-high resolution mapping of the clustered cleavage sites. *Nucl. Acids Res.* **11**, 7899-7910.

Shen, C.-K. J.; and Maniatis, T. (1980). Tissue-specific DNA methylation in a cluster of rabbit B-like globin genes. *Proc. Natl. Acad. Sci. (U.S.A.)* **77**, 6634-6638.

Shepard, J. H.; Mulvihill, E. R.; Thomas, P. S.; and Palmiter, R. D. (1980). Commitment of chick oviduct tubular gland cells to produce ovalbumin mRNA during hormonal withdrawal and restimulation. *J. Cell. Biol.* **87**, 142-151.

Shermoen, A. W.; and Beckendorf, S. K. (1982). A complex of interacting DNAase I-hypersensitive sites near the Drosphila glue protein Sgs4. *Cell* **29**, 601-607.

Simon, D.; Stuhlmann, H.; Jahner, D.; Wagner, H.; Werner, E.; and Jaenisch, R. (1983). Retrovirus genous mehylated by mammalian but not bacterial methylase are non-infectious. *Nature* **304**, 275-277.

Sinden, R. R.; Carlson, J. O.; and Pettijohn, D. E. (1980). Torsional tension in the DNA double helix measured with trimethylpsoralen in living E. coli cells: and analagous measurements in insect and human cells. *Cell* **21**, 773-783.

Smith, R. D.; and Yu, J. (1984). Alterations in globin gene chromatin conformation during muring erythroleukemia cell differentiation. *J. Biol. Chem.* **259**, 4609-4615.

Sollner-Webb, B.; and Felsenfeld, G. (1975). A comparison of the digestion of nuclei and chromatin by staphylococcal nuclease. *Biochemistry* **14**, 2915-2920.

Spadafora, C.; and Crippa, M. (1984). Compact structure of ribosomal chromatin in *Xenopus laevis*. *Nucl. Acids Res.* **12**, 2691-2704.

Spandidos, D. A.; and Paul, J. (1982). Transfer of human globin genes to erythroleukemic mouse cells. *EMBO J.* **1**, 15-20.

Stalder, J.; Seeback, T.; and Braun, R. (1978). Degradation of the ribosomal genes by DNase I in Physarum polycephalum. *Eur. J., Biochem.* **90**, 391-395.

Stalder, J.; Groudine, M.; Dodgson, J. B.; Engel, J. D.; and Weintraub, H. (1980a). Hb switching in chickens. *Cell* **19**, 973-980.

Stalder, J.; Larsen, A.; Engel, J. D.; Dolan, M.; Groudine, M.; and Weintraub, H. (1980b). Tissue-specific DNA cleavages in the globin chromatin domain introduced by DNase I. *Cell* **20**, 451-460.

Stein, R.; Razin, A.; and Cedar, H. (1982). *In vitro* methylation of the hamster adenine phosphoribosyl transferase gene inhibits its expression in mouse L-cells. *Proc. Natl. Acad. Sci. (U.S.A.)*, **79**, 3418-3422.

Stein, R.; Sciaky-Gallili, N.; Razin, A.; and Cedar, H. (1983). Pattern of methylation of two genes coding for housekeeping functions. *Proc. Natl. Acad. Sci. (U.S.A.)* **80**, 2422-2426.

Storb, U.; Arp, B.; and Wilson, R. (1979). The switch region associated with immunoglobulin C genes is DNase I hypersensitive in T lymphocytes. *Nature* **294**, 90-92.

Storb, U.; Wilson, R.; Selsing, E.; and Walfield, A. (1981). Rearranged and germline immunoglobin kappa genes: different states of DNase I sensitivity of constant kappa genes in immunocompetent and nonimmune cells. *Biochemistry* **20**:990-996.

Strauss, F.; and Prunell, A. (1982). Nucleosome spacing in rat liver chromatin. A study with exonuclease III. *Nucl. Acids Res.* **10**, 2275-2293.

Swaneck, G. E.; Nordstrom, J. L.; Kreuzaler, F.; Tsai, M.-J.; and O'Malley, B. W. (1979). Effect of estrogen on gene expression in chicken oviduct. Evidence for transcriptional control of oval-bumin gene. *Proc. Natl. Acad. Sci. (U.S.A.)* **76**, 1049-1053.

Sweet, R. W.; Chao, M. V.; and Axel, R. (1982). The structure of the thymidine kinase gene promoter: nuclease hypersensitivity correlates with expression. *Cell* **31**, 347-353.

Swerdlow, P. S.; and Varshavsky, A. (1983). Affinity of HMG 17 for a mononucleosome is not influenced by the presence of ubiquitin-H2A semihistone but strongly depends on DNA fragment size. *Nucl. Acids Res.* **11**, 387-401.

Talkington, C. A.; and Leder, P. (1982). Rescuing the *in vitro* function of a globin pseudogene promoter. *Nature* **298**, 192-195.

Tanaka, K.; Appella, E.; and Jay, G. (1983). Developmental activation of the H-2K gene is correlated with an increase in DNA methylation. *Cell* **35**, 457-465.

Tilghman, S. M.; and Belayew, A. (1982). Transcriptional control of the murine albumin/α-fetoprotein locus during development. *Proc. Natl. Acad. Sci. (U.S.A.)* **79**, 5254-5257.

Tjian, R. (1978). The binding site on SV40 for a T antigen-related protein. *Cell* **13**, 165-179.

Travers, A. (1983). Protein contacts for promoter location in eukaryotes. *Nature* **303**, 755.

Treisman, R.; Green, M. R.; and Maniatis, T. (1983a). Cis and trans activation of globin gene transcription in transient assays. *Proc. Natl. Acad. Sci. (U.S.A.)* **80**, 7428-7432.

Treisman, R.; Orkin, S. H.; and Maniatis, T. (1983b). Specific transcription and RNA splicing defects in five cloned B-thalassemia genes. *Nature* **302**, 591-596.

Tuan, D.; and London, I. M. (1984). Mapping of DNase I-hypersensitive sites in the upstream DNA of human embryonic Σ-globin gene in K562 leukemia cells. *Proc. Natl. Acad. Sci. (U.S.A.)* **81**, 2718-2722.

van der Ploeg, L. H. T.; and Flavell, R. A. (1980). DNA methylation in the human $\gamma\lambda B$-globin locus in erythroid and nonerythroid tissues. *Cell* **19**, 947-958.

Vardimon, L.; Kressman, A.; Cedar, H.; Machler, M.; and Doerfler, W. (1982). Expression of a cloned adenovirus gene is unhibited by in vitro methylation. *Proc. Natl. Acad. Sci. (U.S.A.)* **79**, 1073-1077.

Varshavsky, A. J.; Sundin, O.; and Bohn, M. (1979). A stretch of "late" SV40 viral DNA about 400 bp long which includes the origin of replication is specifically exposed in SV40 minichromosomes. *Cell* **16**, 453-466.

Villeponteau, B.; Landes, G. M.; Pankratz, M. J.; and Martinson, H. G. (1982). The chicken B-globin gene region: Delineation of transcription units and developmental regulation of interspersed DNA repeats. *J. Biol. Chem.* **257**, 11015-11023.

Villeponteau, B.; Lundell, M.; and Martinson, H. G. (1984). Torisonal stress promotes the DNase I sensitivity of active genes. *Cell* **39,** 469-478.

Volloch, V.; and Housman, D. (1981). Stability of globin mRNA in terminally differentiating murine erythroleukemia cells. *Cell* **23,** 509-514.

Waalwijik, C.; and Flavell, R. A. (1978). MspI, and isoschizoner of HpaII which cleaves both unmethylated and methylated HpaII sites. *Nucl. Acids Res.* **5,** 3231-3236.

Weber, J.; Jelinek, W.; and Darnell, J. E., Jr. (1977). The definition of large viral transcription unit late in Ad 2 infection of HeLa cells: mapping of nascent RNA molecules labeled in isolated nuclei. *Cell* **10,** 611-616.

Weintraub, H. (1983). A dominant role for DNA secondary structure in forming hypersensitive structures in chromatin. *Cell* **32,** 1191-1203.

Weintraub, H.; and Groudine, M. (1976). Chromosomal subunits in active genes have an altered conformation. *Science* **193,** 848-856.

Weintraub, H.; Worcel, A.; and Alberts, B. (1976). A model for chromatin based upon two symmetrically paired half-nucleosomes. *Cell* **9,** 409-417.

Weintraub, H.; Larsen, A.; and Groudine, M. (1981). α-globin gene switching during the development of chicken embryos: expression and chromosome structure. *Cell* **24,** 333-344.

Weisbrod, S. T. (1982a). Active chromatin. *Nature* **297,** 289-295.

Weisbrod, S. T. (1982b). Properties of active nucleosomes as revealed by HMG 14 and 17 chromatography. *Nucl. Acids Res.* **10,** 2017-2042.

Weisbrod, S.; and Weintraub, H. (1979). Isolation of a subclass of nuclear proteins responsible for conferring a DNase I-sensitive structure on globin chromatin. *Proc. Natl. Acad. Sci. (U.S.A.)* **76,** 630-634.

Weisbrod, S.; and Weintraub, H. (1981). Isolation of actively transcribed nucleosomes using immobilized HMG 14 and 17 and an analysis of α-globin chromatin. *Cell* **23,** 391-400.

Weisbrod, S.; Groudine, M.; and Weintraub, H. (1980). Interaction of HMG 14 and 17 with actively transcribed genes. *Cell* **14,** 289-301.

Weischet, W. O.; Glotov, B. O.; Schnell, H.; and Zachau, H. G. (1982). Differences in the nuclease sensitivity between the two alleles of the immunoglobulin kappa light chain genes in mouse liver and myeloma nuclei. *Nucl. Acids Res.* **12,** 3627-3645.

Wigler, M.; Sweet, R.; Sim, G. K.; Wold, B.; Pellicer, A.; Lacy, E.; Maniatis, T.; Silverstein, S.; and Axel, R. (1979). Transformation of mammalian cells with genes from procaryotes and encaryotes. *Cell* **16,** 777-785.

Wilks, A. F.; Cozens, P. J.; Mattaj, I. W.; and Jost, J.-P. (1982). Estrogen induces a demethylation at the 5' end region of the chicken vitellogenin gene. *Proc. Natl. Acad. Sci. (U.S.A.)* **79,** 4252-4255.

Wittig, S.; and Wittig, B. (1982). Function of a tRNA gene promoter depends on nucleosome position. *Nature* **297,** 31-38.

Wong, Y.-C.; O'Connell, P.; Rosbash, M.; and Elgin, S. C. R. (1981). DNase I hypersensitive sites of the chromatin for Drosophila melanogaster ribosomal protein 49 gene. *Nucl. Acids Res.* **9,** 6749-6762.

Wood, W. I.; and Felsenfeld, G. (1982). Chromatin structure of the chicken *B*-globin gene region: sensitivity to DNase I, micrococcal nuclease, and DNase II. *J. Biol. Chem.* **257,** 7730-7736.

Worcel, A.; Gargiulo, G.; Jessee, B.; Udvardy, A.; Louis, C.; and Schedl, P. (1983). Chromatin fine structure of the histone gene complex of *Drosophila melanogaster. Nucl. Acids Res.* **11,** 421-439.

Wright, S.; deBoer, E.; Grosveld, F. G.; and Flavell, R. A. (1983). Regulated expression of the human *B*-globin gene family in murine erythroleukemia cells. *Nature* (London) **305,** 333-336.

Wright, S.; Rosenthal, A.; Flavell, R.; and Grosveld, F. (1984). DNA sequences required for regulated expression of *B*-globin genes in murine erythroleukemia cells. *Cell* **38,** 265-273.

Wu, C. (1980). The 5' ends of Drosophila heat shock genes in chromatin are hypersensitive to DNAase I. *Nature* **286,** 854–860.

Wu, C. (1984). Two protein-binding sites in chromatin implicated in the activation of heat-shock genes. *Nature* **309,** 229–234.

Wu, C.; and Gilbert, W. (1981). Tissue-specific exposure of chromatin structure at the 5' terminus of the rat preproinsulin II gene. *Proc. Natl. Acad. Sci. (U.S.A.)* **78,** 1577–1580.

Wu, C.; Wong, Y.-C.; and Elgin, S. C. R. (1979). The chromatin structure of specific genes: II. Disruption of chromatin structure during gene activity. *Cell* **16,** 807–814.

Young, N. S.; Benz, E. J., Jr.; Kantor, J. A.; Kretschmer, P.; and Nienhuis, A. W. (1978). Hemoglobin switching in sheep: only the γ gene is in the active conformation in fetal liver but all the B and γ genes are in the active conformation in bone marrow. *Proc. Natl. Acad. Sci. (U.S.A.)* **75,** 5884–5888.

Young, R. A.; Hagenbuchle, O.; and Schibler, U. (1981). A single mouse α-amylase gene specifies two different tissue-specific mRNAs. *Cell* **23,** 451–458.

3

Organization of Mitotic Chromosomes

Kenneth W. Adolph
University of Minnesota Medical School
Minneapolis, Minnesota

COMPONENTS OF METAPHASE CHROMOSOMES

DNA: Each Chromosome Contains Essentially a Single DNA Molecule

A variety of compelling evidence supports the idea of one chromatin fiber per chromatid (the "unineme hypothesis"). Such evidence includes the direct detection of chromosome-length DNA molecules, the semiconservative segregation of labeled DNA strands, and the existence of linear genetic maps.

That chromosome-sized molecules could exist was first suggested by autoradiography. Cairns (1966) and Huberman and Riggs (1966) were able to measure molecules of mammalian DNA up to 2 mm in length. Viscometric analysis of DNA from *Drosophila* cells showed that chromosome-length DNA molecules were present (Kavenoff and Zimm, 1973), and direct detection by autoradiography revealed molecules 1.2 cm long (Kavenoff et al., 1974). Autoradiography of intact chromosomes demonstrated the semiconservative distribution of [^{3}H]thymidine label during mitosis (Taylor et al., 1957). The semi-conservative distribution of label in chromosomes is most simply explained as reflecting the semi-conservative replication of the DNA double helix. Thus, the results provide strong evidence against chromosomes containing multiple, identical DNA molecules. The presence of one DNA strand for each of the loops of mammalian lampbrush chromosomes (see chap. 5) was suggested by kinetic analysis of breakage of the chromosomes with deoxyribonuclease (Gall, 1963). Irradiation of chromosomes with X-rays showed a linear relationship between the number of breaks and the dose, which is most readily explained if the target consists of a single DNA stand (Waldren and Johnson, 1974). Genetic phenomena such as sister chromatid exchanges, translocations, and deletions are also most easily accounted for if a single DNA molecule is broken and rejoined. Furthermore, the physical maps (G-banding patterns) and genetic maps of human chromosomes appear to be colinear (Cook and Hamerton, 1979), which again must be due to a single chromatin fiber extending along the entire chromatid length.

Histones: The Nucleosome Structure of Interphase Chromatin is Preserved in Metaphase

In the interphase nucleus, histones are packaged with DNA through the formation of nucleosomes. The DNA is compacted by a factor of seven by coiling the DNA double helix around the nucleosome core, which consists of two each of histones H2A, H2B, H3, and H4. A further sixfold contraction of the length of DNA occurs by creation of the basic 300 Å chromatin fibers. These fibers result from the helical coiling of the 100 Å beads-on-a-string nucleosome filament into a superhelix or solenoid. Are nucleosomes and solenoids conserved in metaphase? Experimental evidence indicates that these structures are not only preserved, but are essentially unchanged in metaphase chromosomes.

Compton et al. (1976) compared the subunit structure of interphase chromatin and mitotic chromosomes from Chinese hamster ovary cells. Digestion of both nuclei and chromosomes produced DNA fragments that were integral multiples of a repeating monomer unit of approximately 177 base pairs. Thus the nucleosome substructure of interphase chromatin appeared to be unaltered in metaphase, a result that was supported by electron microscopic observations. Linear chains of tightly packed spheres (nucleosomes), 120 Å in diameter, were seen for partially unravelled chromosomes.

Low angle X-ray diffraction studies of HeLa metaphase chromosomes and isolated nuclei were undertaken to discover whether major changes in chromatin structure occur during the cell cycle (Langmore and Paulson, 1983; Paulson and Langmore, 1983). Only the packing of the chromatin fibers was found to be affected by the transition to metaphase. Both nucleosome arrangement within chromatin fibers and the internal structure of nucleosomes were unchanged within the resolution of the technique. Of particular interest was the observation that fiber structure in metaphase was not influenced by the level of phosphorylation of histones H1 and H3. The substantial increase in H1 phosphorylation during mitosis had led to the proposal that this phosphorylation actually triggers chromosome condensation by modifying histone-DNA interactions.

Nonhistone Proteins: Metaphase Chromosomes Possess a Complex Nonhistone Composition

Although the interactions of histones in creating the 300 Å chromatin fibers are understood in considerable detail, the roles of the nonhistone proteins that are components of metaphase chromosomes remain largely unknown. Hundreds of nonhistones are found to be associated with metaphase chromosomes using two-dimensional polyacrylamide gel electrophoresis. Many of these may have specific structural roles in the condensation of metaphase chromosomes and in maintaining the characteristic chromosome morphology. Many others may not have any roles that are specific to metaphase, and may be simply carried along as components of chromatin fibers during chromosome condensation. Understanding the broader pattern of structural interactions of nonhistones in metaphase remains a formidable task.

An initial approach to understanding the nature and significance of nonhistone interactions has been to consider mitosis as part of the cell cycle and to investigate changes in the patterns of nonhistone synthesis and modification from interphase into metaphase. These studies, to be described, used synchronized cultures of HeLa cells, usually prepared by the double thymidine block procedure, and, in addition, metaphase chromosomes were isolated from colchicine-arrested cells. Phosphorylation and ADP-ribosylation were the post-translational modifications that were investigated, since previous studies had demonstrated that these were significant modifications of chromatin-associated proteins.

Synthesis of HeLa cell nonhistones from interphase into metaphase was monitored by using a radioactive label of [^{35}S]methionine (Adolph and Phelps, 1982). Synchronized populations of cells were pulse-labeled with the isotope for 1.0 hr periods before nuclei were isolated and chromatin was obtained by brief digestion with micrococcal nuclease. The chromatin was fractionated on sucrose density gradients, and both one-dimensional and two-dimensional gel electrophoresis were employed to separate the labeled nonhistones. Figure 3-1 shows the typical appearance of autoradiograms of

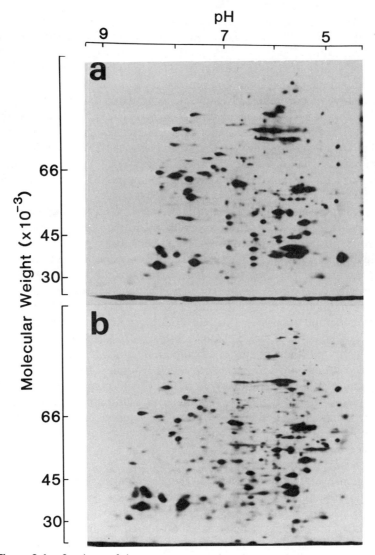

Figure 3-1. Synthesis of chromatin-associated nonhistone proteins from synchronized HeLa cells. Cells were labeled with [^{35}S] methionine at (a) 0–1 hr and (b) 12–13 hr after release from the G1/S boundary. Proteins were separated by two-dimensional polyacrylamide gel electrophoresis., The first dimension separation used the non-equilibrium pH gradient procedure, while the second dimension used electrophoresis in 8% SDS-polyacrylamide gels. Proteins that incorporated radioactive label were detected by fluorography. *(From Adolph and Phelps, 1982, p. 9088)*

2-D gels of chromatin nonhistones labeled at 0-1 hr and 12-13 hr after release of cells from the G1/S boundary. The period from 12-13 hr is close to mitosis, while the 0-1 hr period represents early S phase (the DNA synthesis phase). Examining the autoradiograms reveals significant differences due to the cessation of synthesis of several proteins at 12-13 hr and an increase in the radioactivity incorporated by many minor proteins. However, the proportion of new proteins synthesized close to mitosis is not large, so that the pattern near mitosis resembles that for early S phase. The same conclusion is true when comparing the pattern close to mitosis with the 4-5 hr, 8-9 hr, and 16-17 hr patterns. The 4-5 hr pattern is near the peak of S phase, and so there also does not appear to be a strong correlation between DNA synthesis and changes in nonhistone synthesis.

The results for metaphase chromosomes could be misleading, however, since only about 25% of synchronized cells are in mitosis at the time that corresponds to the peak of mitosis. This is because individual cells have individual cell cycle lengths, and because the degree of synchrony is not complete. Furthermore, metaphase chromosomes are more than condensed chromatin fibers, and have a large ratio of nonhistones to histones. A distinct pattern for metaphase chromosomes could result from the specific synthesis of certain metaphase proteins in late G2 or prophase. To examine this possibility, radioactively labeled chromosomes were isolated from metaphase-arrested cells, and their nonhistone components separated on 2-D gels. It was indeed found that the pattern of labeled nonhistones of metaphase chromosomes differed considerably from that of interphase chromatin. But it was not possible to determine whether the additional nonhistones are synthesized immediately preceding mitosis or whether their synthesis follows the pattern of chromatin nonhistones in not demonstrating a strong correlation with the cell cycle.

Metaphase chromosomes appear to be more abundant in specific nonhistones than are whole nuclei, since 2-D gels have shown that the pattern for nuclei is quite similar to that for interphase chromatin (Adolph and Phelps, 1982). Thus, only a minority of nuclear proteins constitute the protein framework of the nucleus, which can be considered a separate substructure from the chromatin fibers. The proteins of the nuclear framework can be isolated by digesting DNA with nuclease and extracting histones and many other nuclear proteins with a high concentration of salt (2M NaCl) or polyanions. A "nuclear scaffold" is thereby produced, which has prominent proteins (the "lamins") with molecular weights of 60,000-70,000 daltons. The major proteins of the nuclear scaffold were directly exised from 2-D gels for scintillation counting. As for chromatin-associated nonhistones, changes in the synthesis of scaffold proteins were not confined to any particular period in the cell cycle and were not major, differing in measured radioactivity by about twofold at most.

Summarizing the results on nonhistone synthesis, more than 300 nonhistone species that are components of chromatin could be identified by 2-D gel electrophoresis. But large scale alterations in the patterns of synthesis were not found to be associated with mitosis. It remains possible, however, that the synthesis of a few special proteins, immediately preceding mitosis, may have a critical role in chromosome condensation.

If large scale changes in nonhistone synthesis are not crucial for mitotic chromosome condensation, perhaps post-translational modifications of nonhistones are significant. Phosphorylation of nonhistones during the HeLa cell cycle was therefore investigated (Song and Adolph, 1983b). Previous studies of histone phosphorylation had established that a substantial increase in H1 phosphorylation is associated with mitosis, and it was proposed that this phosphorylation is actually responsible for chromosome condensation. To study nonhistone phosphorylation, synchronized cells were labeled in vivo by adding [^{32}P] orthophosphate to the culture medium. Nuclei and chromatin were then isolated, and the ^{32}P-labeled nonhistones were separated on SDS-polyacrylamide gels (Fig. 3-2). Densitometry of autoradiograms and stained gels showed that the highest level of specific radioactivity (^{32}P cpm incorporated/μg protein) occurred in mid-S phase for eight chosen phosphorylated nonhistones. These measurements confirmed observations of autoradiograms, which revealed that all observable nonhistones are most highly phosphorylated in mid-S phase (Fig. 3-2). Close to mitosis the level of nonhistone phosphorylation is substantially less. A similar protocol of in vivo labeling was also followed for isolated chromosomes prepared from metaphase-arrested cells. Autoradiograms demonstrated even more clearly than for chromatin that a dephosphorylation of nonhistones is associated with mitosis. For S phase chromatin, 93% of the ^{32}P radioactivity in proteins (excluding core histones) was in nonhistones, while only 7% was in histones H1B plus H1A. But the situation had reversed for metaphase chromosomes, with 28% of the ^{32}P in nonhistones and 72% in histones H1B plus H1A.

Further analysis of the nature of the phosphorylation showed that for eight selected nonhistones ranging in molecular weight from 36,000 to 138,000 daltons, only serine residues were phosphorylated. Neither threonine nor tyrosine incorporated the radioactive phosphate label. The number of phosphorylated residues per protein molecule was characterized by isoelectric focusing for one prominent phosphoprotein (protein 55K). This protein was seen to be mono- or di-phosphorylated in metaphase, predominantly di-phosphorylated in unsynchronized cells, and di-or tri-phosphorylated in mid-S phase.

Proteins of the nuclear protein framework (the nuclear scaffold) were also found to be phosphoproteins. In addition to the lamins, a protein of 119,000 daltons was found to be a major acceptor of label.

It is therefore seen that phosphorylation of nonhistones is associated with the cell cycle to a considerably greater degree than is the synthesis of nonhistones. But the association is primarily with S phase, and not with mitosis. Indeed, a general dephosphorylation of metaphase nonhistones was discovered. This dephosphorylation could be functionally significant, but it is more likely that this modification of nonhistones does not have central role in chromosome condensation.

ADP-ribosylation is a post-translational modification that has been more recently appreciated as important for chromatin structure (Hilz and Stone, 1976; Hayaishi and Ueda, 1977; Purnell et al., 1980). Histone H1 as well as the core histones are primary targets for modification (Adamietz et al., 1978; Wong et al., 1983), while poly(ADP-ribose) polymerase is tightly bound to chromatin and thus may be considered a nonhistone (Jump and Smulson, 1980; Jump et al., 1979). A possible function for ADP-ribosylation in mitosis

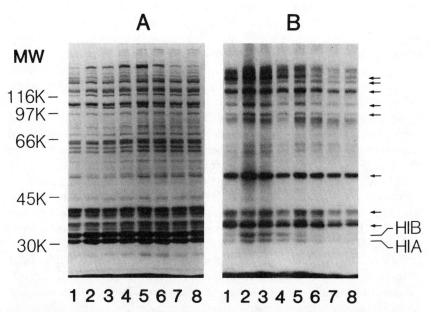

Figure 3-2. Cell cycle variations in phosphorylation of chromatin-associated nonhistones. Synchronized HeLa cells were labeled with [^{32}P] orthophosphate for 1.0 hr at intervals of 2 or 4 hr after release from G1/S. Chromatin was prepared by digestion of isolated nuclei with micrococcal nuclease, and chromatin-associated nonhistones were resolved by electrophoresis in 8% SDS-polyacrylamide gels. The periods of labeling after release of cells from G1/S were: (1) 0–1 hr; (2) 4–5 hr; (3) 6–7 hr; (4) 8–9 hr; (5) 10–11 hr; (6) 12–13 hr; (7) 16–17 hr; and (8) 20–21 hr. (A) Staining pattern, using Coomassie brilliant blue, of the chromatin nonhistones. (B) Autoradiogram revealing the ^{32}P-labeled proteins. Eight major phosphorylated nonhistones, selected for more detailed characterization, are indicated at the right by arrows. (From Song and Adolph, 1983b, p. 3310)

was suggested by the observation that the highest level of modification during the cell cycle was during metaphase (Tanuma and Kanai, 1982); Holtlund et al., 1983). The activity of poly (ADP-ribose) polymerase is associated with cellular events that involve DNA strand breaks, including DNA excision repair, gene expression and cell differentiation (Farzaneh et al., 1982; Althaus et al., 1982; Johnstone and Williams, 1982).

ADP-ribosylation of nonhistone proteins during the HeLa cycle has been investigated (Song and Adolph, 1983a). The immediate precursor for the modification, NAD, is bulky and does not readily penetrate the cell to reach the nuclear nonhistones. Therefore, the experiments were carried out by labeling isolated nuclei with [^{3}H] adenosine, which has the disadvantage that this label is not the direct precursor and so may not be attached to protein as poly (ADP-ribose). Additionally, since most label is incorporated into nucleic acid and very little is linked to protein, autoradiography frequently requires months to adequately expose X-ray film.

Incubating isolated interphase nuclei with [^{32}P]NAD and separating labeled proteins on 2-D gels showed that ADP-ribosylation is a major post-translational modification of nuclear nonhistones, affecting more than 100 protein species. The primary acceptors of label were the lamins, the structural proteins of the peripheral nuclear lamina, and a 116,000 dalton species that appears to be poly (ADP-ribose) polymerase. But many other proteins, possessing a broad spectrum of molecular weights and isoelectric points, were also labeled. A similar pattern of modified nonhistones was found for chromatin (Fig. 3-3). The primary difference was that the lamins were no longer present with chromatin. A much simpler distribution of ADP-ribosylated proteins was seen for autoradiograms of nuclear scaffold proteins. Besides the lamins, the only other significant acceptor of label was the species identified as poly (ADP-ribose) polymerase.

The most striking result of this study was found for metaphase chromosome nonhistones. Although hundreds of nonhistones are components of metaphase chromosomes, autoradiograms of 2-D gels of isolated metaphase chromosomes showed only a single major acceptor of label from [^{32}P]NAD. This is the 116,000 dalton species which appears to be the automodified poly (ADP-ribose) polymerase. In these experiments, mitotic cells were permeabilized by being resuspended in hypotonic buffer containing non-ionic detergent (0.1% NP40). Thus, for most nonhistones there is a substantial decrease in ADP-ribosylation of nonhistones during mitosis. Any increase in the overall level of modification during metaphase (Tanuma and Kanai, 1982; Holtlund et al., 1983) must therefore be exclusively due to an increase in the modification of the 116,000 dalton protein. The decrease in nonhistone modification may result for steric reasons, because of the extreme degree of compaction of metaphase chromosomes, or may be under more specific control.

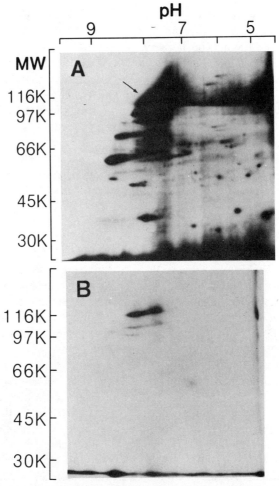

Figure 3-3. ADP-ribosylated proteins of HeLa interphase chromatin and metaphase chromosomes separated on two-dimensional gels. The direct precursor, [^{32}P]NAD, was used to label isolated interphase nuclei and permeabilized mitotic cells. Chromatin was obtained from nuclei by digestion with micrococcal nuclease, while metaphase chromosomes were isolated and purified on Metrizamide density gradients. *(A)* Autoradiogram displaying ADP-ribosylated nonhistones of interphase chromatin. The arrow indicates the major acceptor of label. *(B)* Autoradiogram showing radioactively labeled proteins of metaphase chromosomes. For both autoradiograms, electrophoresis used the nonequilibrium pH gradient technique in the first dimension and used 8% SDS-polyacrylamide gels in the second dimension. *(From Song and Adolph, 1983a, pp. 942–943; copyright © 1983 by Academic Press)*

The same general results were found by using in vivo labeling with [^{3}H] adenosine. Autoradiograms of 2-D gels of nuclei and chromatin showed a large number of modified nonhistones, again demonstrating that ADP-ribosylation is not restricted to only a few special proteins. And the fundamentally distinct pattern for metaphase chromosomes, with only a single significant acceptor of [^{3}H] adenosine label, was also observed. This establishes that the change discovered with [^{32}P]NAD is not the result of the in vitro labeling protocol, but actually reflects a physiological difference between interphase and metaphase. However, the autoradiogram patterns for the large number of nuclear proteins labeled with [^{32}P]NAD and [^{3}H] adenosine are not identical, which may show that the labeling conditions do have some effect.

Unlike the situation with phosphorylation, the function of ADP-ribosylation of nonhistones and other proteins is, in general, not clear. Thus, it is quite difficult to speculate on the significance of this nonhistone modification for changes in chromosome organization. But the demonstration of a fundamental difference between interphase and metaphase may indicate a structural role associated with metaphase chromosome condensation.

Comparison of 2-D gels of proteins labeled with [^{35}S] methionine showed that the patterns for nuclei and metaphase chromosomes are very different. The question may therefore be posed as to the extent of nonhistone conservation from interphase to metaphase. Immunological procedures ("immunoblotting") have therefore been applied (Adolph, 1984). Antibodies were raised in rabbits to the proteins of interphase HeLa cell chromatin. Proteins of nuclei, isolated metaphase chromosomes, interphase chromatin, and chromatin from metaphase chromosomes were separated on SDS-polyacrylamide gels and electrophoretically transferred to diazophenylthioether (DPT)-paper. The paper was then covered with antiserum specific to interphase chromatin. ^{125}I-labeled protein A was used to detect the bound antibodies. An alternate procedure was to directly overlay polyacrylamide gels with antiserum, and thus avoid transferring the proteins to an immobilizing support. The degree of conservation of nonhistone antigenic determinants from interphase to metaphase could then be measured using densitometry of autoradiograms and stained gels. It was found that over 90% of interphase chromatin nonhistones are conserved in metaphase. Employing 2-D gels demonstrated that at least 80 distinct nonhistone species are conserved. Therefore, beneath the appearance of a very different pattern of [^{35}S] methionine labeled species in metaphase, which is due to the addition of unique metaphase nonhistones, most species that are bound to chromatin in interphase remain bound in metaphase. A quarter to a third of HeLa chromatin nonhistones were also conserved as components of chromatin from other species (chicken erythrocytes, Novikoff rat hepatoma cells, mouse L cells).

Isolation of Metaphase Chromosomes

The ability to perform meaningful biochemical and structural experiments with metaphase chromosomes requires techniques for the isolation of chromosomes that are as physiological as possible. Harsh isolation conditions that modify the native macromolecular interactions are unacceptable. Special care must be taken since unphysiological procedures may produce a good yield of clean particles with the expected metaphase chromosome morphology, but with altered biochemical properties. In addition to being physiological, the isolation conditions must also allow chromosomes to be readily purified from contaminating material, such as cytoskeletal debris. Obtaining large quantities of mitotic cells has also been a limiting factor, although the use of tissue culture cells and the availability of drugs such as colchicine to arrest cells in mitosis has mitigated this difficulty. However, a primary hindrance to progress in biochemical and structural studies of isolated chromosomes has been in the lack of techniques for chromosome isolation.

A useful procedure for chromosome isolation was presented by Wray and Stubblefield (1970). This method employs an isolation buffer containing 1.0 M hexylene glycol (2-methyl-2,4-pentanediol), 0.5 mM $CaCl_2$, 0.1 mM PIPES pH 6.5, and chromosomes are released by shearing mitotic cells through a syringe needle. A variation of this procedure, which uses a pH of 10.5 to minimize degradation of nucleic acids, has also been reported (Wray et al., 1972). The major advantage of the procedure is that a good yield of chromosomes is obtained with little cytoskeletal contamination and with a well-defined chromosome morphology. Furthermore, acid conditions, which were included in earlier procedures to stabilize chromosomes during their isolation, are avoided. However, the use of an organic solvent in preparing chromosomes is questionable, especially if biochemical experiments are to be carried out.

New methods were developed for the isolation of metaphase chromosomes from HeLa cells that use entirely aqueous buffers (Adolph, 1980). Either divalent cations (Mg^{2+}, Ca^{2+}) or polyamines (spermine/spermidine) are utilized to stabilize chromosomes in a condensed state during isolation. The concentration of monovalent salt (NaCl) is also important, and was adjusted to 50-80 mM to optimally reduce both the number of stretched chromosomes and the amount of cellular debris. Chromosomes were purified either by pelleting through sucrose or by sedimentation into preformed density gradients of Metrizamide, a nonionic medium.

Variations of preparative techniques that employ hexylene glycol, divalent cations and polyamines were examined by Lewis and Laemmli (1982) in their investigation of the role of metalloprotein interactions in stabilizing histone-depleted chromosomes. Density gradients of Percoll were found to be highly effective in reducing cytoskeletal contamination. Their technique involving hexylene glycol was a modification of the Wray and Stubblefield (1970) procedure, while the use of polyamines was adapted from Blumenthal et al. (1979).

STRUCTURAL ORGANIZATION OF INTACT CHROMOSOMES

Metaphase Chromosomes: These Highly Organized Structures Are Representative of Condensed States of Chromatin

DNA, histones, and nonhistone proteins are assembled in the eukaryotic cell to create chromosomes, and these packages of genetic information are in their most condensed state during metaphase in mitosis. Evidence that has accumulated over decades of research demonstrates that metaphase chromosomes have a well-defined and constant substructure. The persistence of the same sizes and shapes of chromosomes from cell division to cell division is an indication that chromosome structure is constant, and is maintained, although in a diffuse state, through interphase. The production of reproducible and characteristic patterns of bands known as G-, C-, and Q-bands is direct evidence that not only do chromosomes have an overall morphology that is well-defined, but the same is true of the detailed chromosomal substructure. The number of distinct G-bands that can be reproducibly recognized for human chromosomes is now at least 1,256 for prophase chromosomes, though these coalesce by linear contraction of chromosomes into 320 bands in mid-metaphase (Yunis, 1976).

Evidence also suggests that metaphase chromosomes are not unique structures, but that the same way of packaging chromatin fibers is found with other chromosome types. As will be discussed in detail in this review, the chromatin fibers of metaphase chromosomes appear to be predominantly organized as radial arrays of loops. Metaphase chromosome substructure can be considered the basis of organization for such diverse chromosome types as mitotic prophase chromosomes, meiotic chromosomes, lampbrush chromosomes, interphase chromosomes, and polytene chromosomes. This conclusion follows from the structural parallels that are observable in particular cases. For instance, meiotic chromosomes reveal the same G- and Q-banding patterns as do metaphase chromosomes (Caspersson et al., 1971).

Thin Sectioning and Transmission Electron Microscopy

Electron microscopy is the most valuable tool that is available to investigate the structure of mitotic chromosomes. This is because the scale of structural detail that is characterized ranges from tens of angstroms along the nucleosomal chromatin fibers to ten or more microns from one end of a chromatid to the other. Only electron microscopy is able to provide a direct picture of variations in the arrangement of matter over this immense range of dimensions. The two most useful techniques have proven to be transmission electron microscopy of thin sections through chromosomes, and scanning electron microscopy of isolated chromosomes. The distribution of the material of the chromatin fibers across and along the body of the chromatids is revealed by thin sectioning, while details of surface organization of individual chromosomes is shown by scanning electron microscopy.

The primary obstacle that is encountered when attempting to successfully apply these procedures is the large size of each chromosome combined with the dense and intricate arrangement of chromosomal fibers. This is true not only for human mitotic chromosomes (HeLa cells), but also for chromosomes from other higher eukaryotes such as mice (mouse L929 cells) and rats (Novikoff rat hepatoma N1S1-67 cells). Because of this dense arrangement of material, investigations of native chromosomes prepared for electron microscopy under in situ conditions are not particularly informative. Although intense staining makes compact areas discernible in thin sections, individual fibers cannot be resolved. Thus, while electron microscopy has been applied for many years to the problem of chromosome structure, it has only been possible to draw limited conclusions regarding the organization of chromosomal fibers. The earliest applications that yielded some information involved whole-mount electron microscopy (using transmission EM) and scanning EM, and these will be discussed.

Gaining some information about how the DNA and chromatin fibers are arranged requires conditions that depart from those in vivo. The essential result must be that chromatin or DNA fibers are separated and, therefore, distinguishable. But, at the same time, it must be possible to extrapolate and relate the results to the native distribution of fibers. Two extreme approaches have been utilized. Histones and many nonhistones can be completely extracted (see the section on Histone-Depleted Chromosomes and Chromosome Scaffolds). The extraction is carried out under mild conditions such that some remnant of structure remains that can be characterized by electron microscopy (Adolph et al., 1977a; Paulson and Laemmli, 1977). Another approach, and one that will be discussed in detail in this section, involves slightly expanding chromosomes without removing any proteins, so that individual fibers can be resolved. The chromosomes are expanded by adjusting the solution conditions of resuspension buffer and, in particular, by adjusting the concentrations of divalent cations (Mg^{2+}, Ca^{2+}) or monovalent salt (NaCl). This latter approach has been applied to a serial sectioning study of the structure of HeLa metaphase chromosomes (Adolph, 1981). The results were extremely informative in providing tantalizing glimpses of how the chromatin fibers are arranged.

What is the objective of these electron microscopic studies of metaphase chromosome structure? The fundamental aim is to develop a concrete model for the general path of folding of the chromatin fibers. This is a very basic aim, but, even in the mid-1980s, considerable uncertainty and controversy remains. At an initial level, electron microscopy can yield some information about the distribution and orientation of the fibers. Are the fibers uniformly distributed across the chromatids, or do holes exist? Are the fibers primarily longitudinal to the central chromatid axis, or primarily radial? At a higher level would be an understanding of precisely how the fibers progress along the chromatid arms. This is more difficult to achieve, since it requires the ability to trace a fiber for a considerable distance through three dimensions. A degree of sophisticated analysis beyond visual examination of electron micrographs would seem to be required. A description of any regularity or symmetry in fiber packing would be needed. For example, does helical symmetry, which determines the arrangement of nucleosomes in the 300 Å chromatin fibers, also have a primary role in the higher-order structure of assembled chromosomes?

Another objective of electron microscopy is to understand the involvement of nonhistone proteins and to understand the unique regions (centromeres, telomeres) of chromosomes. Mitotic chromosomes are more than uniformly folded fibers, but are complex structures with specialized regions and perhaps with separate structural compartments.

The central approach in a serial sectioning study of HeLa metaphase chromosomes (Adolph, 1981) was to slightly expand chromosomes without removing proteins, so that chromosomal fibers were separated and their general orientation could be observed. This was accomplished by resuspending metaphase-arrested cells in hypotonic buffer containing 1.0-1.5 mM Mg^{2+} and 0.1% NP40. The nonionic detergent NP40 solubilized membraneous material adjacent to the chromosomes. A Mg^{2+} concentration of 1.0-1.5 mM was found to give the most informative results in allowing individual fibers to be distinguished. Two basic types of sections were collected: transverse sections, sections essentially perpendicular to the central chromatid axis, and longitudinal sections, sections parallel to the chromatid axis. Sections were collected serially through substantial portions of chromosomes.

Both transverse and longitudinal sections show that the fibers are primarily organized in a radial distribution (Figs. 3-4, 3-5). That is, the bulk of the chromatin fibers are oriented perpendicular to the central chromatid axes, and apparently loop back at the peripheries of the chromatids. One must say apparently loop back, since it is difficult to follow complete loops. The fibers twist out of the plane of the sections, and the degree of disorder in the micrographs also precludes readily tracing full loops. The radial arrangement of fibers extends uniformly along each chromatid arm. Although this statement is generally true, the appearance of certain micrographs suggests that the detailed organization may be somewhat more complex than a totally uniform array of radial loops.

A fundamental transition can be observed as the sections enter the body of a chromosome. As the sectioning knife first encounters the chromosome (in 1.0-1.5 mM Mg^{2+}), a dot pattern is observed. This pattern of dots must result from the knife cutting across radially oriented fibers, thus providing additional evidence for the radial loop model. As the sections penetrate deeper and deeper into the body of a chromatid, the pattern of dots becomes denser and is then gradually replaced by an array of fibers in the plane of the section characteristic of a longitudinal section. This is again what is to be expected for radially arranged loops of chromatin.

Scanning Electron Microscopy

Scanning EM can be profitably applied to the question of how the chromatin fibers are organized in metaphase chromosomes. The information provided by this technique concerns the surface structure of chromosomes. To gain useful information, it is therefore essential that this surface structure be uncovered. Two procedures have been employed: swelling the chromosomes by adjusting the concentration of divalent cations (Adolph and Kreisman, 1983), and treating (uncoating) chromosomes with trypsin as for G-banding (Harrison et al., 1981, 1982, 1983). Application of both of these procedures has resulted in the striking observation that metaphase chromosomes have a knobby surface structure. These surface protuberances would seem to be related to the underlying radially oriented fibers.

Scanning EM has been used by investigators of chromosome substructure for a number of years (Christenhuss et al., 1967; Scheid and Traut, 1971; Golomb and Bahr, 1971; Cervenka et al., 1973; Golomb, 1976; Daskal et al., 1976; Korf and Diacumakos, 1978; Marsden and Laemmli, 1979). Only recently, however, has the value of the technique for high-resolution studies been appreciated and exploited.

Adjusting the concentration of divalent cations in resuspension buffer to expand the chromosomes and reveal the underlying fiber arrangement proved to be valuable for thin sectioning, and so was also employed for SEM

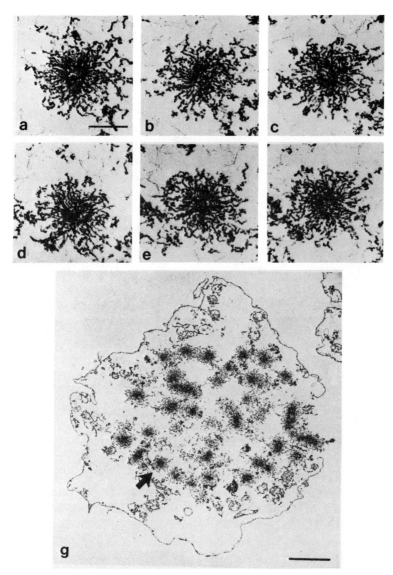

Figure 3-4. Electron micrographs of consecutive, transverse sections through a mitotic HeLa cell chromosome (a–f). The mitotic cell was prepared for transmission electron microscopy in hypotonic buffer containing 0.1% NP40 and 1.5 mM Mg^{2+}. Panel (g) shows a section containing the chromosome in (a–f). An arrow at the lower left points to this chromosome. The bar represents 0.5 μm for (a–f) and 5.0 μm for (g). (From Adolph, 1981, p. 148; copyright © 1981 by European Journal of Cell Biology)

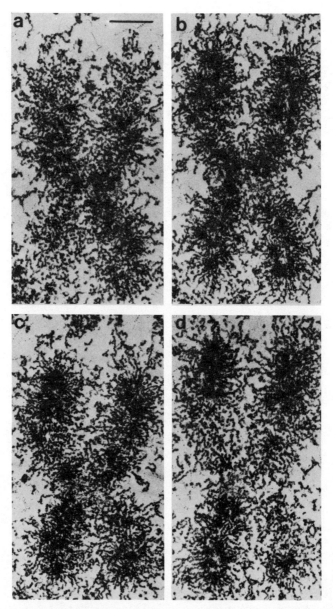

Figure 3-5. Electron micrographs of longitudinal sections through a mitotic HeLa cell chromosome. The hypotonic buffer in which the mitotic cell was resuspended before pro-cessing for electron microscopy contained a Mg^{2+} concentration of 1.5 mM. The bar represents 0.5 μm. *(From Adolph, 1981, p. 149; copyright © 1981 by European Journal of Cell Biology)*

(Adolph and Kreisman, 1983). Isolated chromosomes were obtained, and were processed for SEM by standard techniques that included fixation with glutaraldehyde and osmium tetroxide, dehydration through ethanol or acetone, critical point drying, and coating with gold/palladium. The effect of reducing the Mg^{2+} ion concentration was striking (Fig. 3-6). A relatively smooth surface structure was observed in 5.0 mM Mg^{2+}. This is the concentration of Mg^{2+} in isolation buffer, where chromosomes have similar overall dimensions to their in situ state. Thin sectioning also revealed that, in 5.0 mM Mg^{2+} and for mitotic cells fixed in growth medium, chromosomes are highly condensed with little observable substructure. But for chromosomes prepared in 1.5 mM Mg^{2+}, the change in surface structure is dramatic (Fig. 3-6). Surface protuberances having diameters of almost 700 Å (691 ± 96) are the dominant surface feature. But this knobby surface structure is not a static feature. As the concentration of magnesium is further reduced, the diameter of the protuberances also declines (Fig. 3-7). In 0.5 mM Mg^{2+}, the diameter is about 650 Å (647 ± 76), but only about 350 Å (349 ± 52) in 0.15 mM Mg^{2+}. Decreasing the Mg^{2+} ion concentration to 0.05 mM and below results in a value of approximately 300 Å. This is the thickness of the superhelix of nucleosomes, which constitutes the fundamental chromatin fiber. Indeed, the suggestion of a helical substructure is evident from the scanning electron micrographs, which reveal a disk-like periodicity along the fibers of chromosomes in 0.05 mM Mg^{2+}. This periodicity of approximately 150 Å appears to represent the turns of the helix of nucleosomes, each of which has a diameter of about 110 Å.

These SEM results strongly support the conclusion drawn from thin sectioning that the basic 300 Å chromatin fibers are primarily organized as radial loops. The surface protuberances would represent the peripheral tips of radial loops. Some degree of further compaction or super-twisting must exist in the higher magnesium concentrations (1.5 mM, 0.5 mM), since the diameters are considerably wider (approaching 700 Å) than the fibers (300 Å). In this investigation, the substructure of the protuberances could not be observed in greater detail because of the thickness of the gold/palladium coating. But the value of SEM for studies of this kind is clear, so the approach might be further extended to obtain a more desirable resolution of detail.

Whole-Mount Electron Microscopy

Earlier investigations of metaphase chromosome organization by transmission electron microscopy led to the "folded fiber model" of chromosome structure (DuPraw, 1966, 1970; Gall, 1966; Bahr, 1977). Chromosomes, preserved for electron microscopy by critical-point drying, were seen to possess a dense brushlike appearance. This suggested that chromatin fibers were

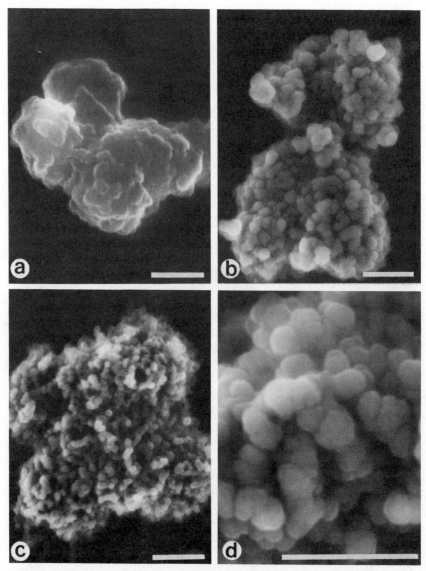

Figure 3-6. Scanning electron micrographs of individual, isolated metaphase chromosomes. The chromosomes were prepared for electron microscopy in Mg^{2+} concentrations of (a) 5.0 mM, (b) 1.5 mM, and (c, d) 0.5 mM. Isolation of chromosomes from metaphase-arrested HeLa cells employed an aqueous buffer containing divalent cations to stabilize chromosome morphology. The bar represents 0.5 μm. *(From Adolph and Kreisman, 1983, p. 157; copyright © 1983 by Academic Press)*

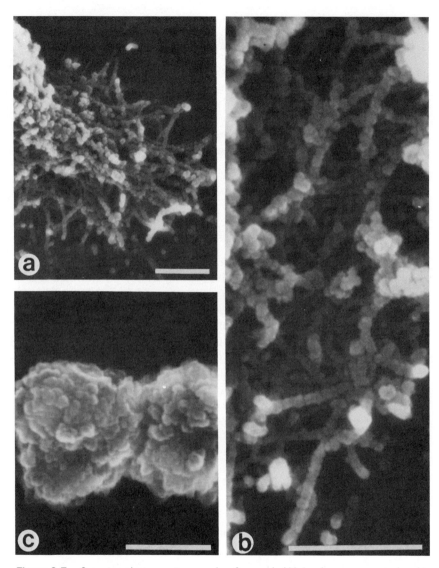

Figure 3-7. Scanning electron micrographs of expanded HeLa chromosomes isolated in buffer containing divalent cations *(a, b)* and of a condensed chromosome isolated in a polyamine-containing buffer *(c)*. The isolated chromosomes in *(a)* and *(b)* were resuspended in a solution containing 0.05 mM Mg^{2+}. The bar represents 0.5 μm. *(From Adolph and Kreisman, 1983, p. 158; copyright © 1983 by Academic Press)*

intricately, but directly, folded into the metaphase chromosome structure without separate levels of organization being present. Central to this view was the "unineme hypothesis" that a metaphase chromatid consists of a single chromatin fiber extending the entire length (Prescott, 1970). The "folded fiber model" remains a fundamentally correct, first approximation. More recent electron microscopy of both intact (Figs. 3-4 through 3-7) and histone-depleted chromosomes supports the notion that the basic chromatin fibers are directly folded into the chromosome morphology. A somewhat different view was put forward by Stubblefield and Wray (1971). Two distinct levels of organization, "epichromatin loops" attached to "core fibers," were proposed.

Organization of Centromeres and Telomeres

Metaphase chromosomes are not uniform cylindrical structures, but consist of specialized regions in addition to the bulk of condensed chromatin. The most prominent of these specialized regions are the centromeres (kinetochores) and telomeres. Located at the primary constriction, kinetochores serve to attach chromosomes to the spindle apparatus. They therefore have a central role in directing chromosomes through mitosis and meiosis. Not only are these regions structurally distinct, but they are also distinct in their sequence of DNA. Similarly, telomeres are not only the positions at which chromosomes end, but have, in addition, special nucleotide sequences.

Mammalian kinetochores have a complex ultrastructure, consisting of dense outer and inner layers separated by a less opaque middle layer to produce a trilaminar disk (Ris and Witt, 1981). The outer dense layer appears to represent chromatin, continuous with the bulk of the chromosome, to which bundles of microtubules are directly attached. Nucleation of kinetochore microtubules occurs within a fibrous corona external to the dense outer layer.

The nucleotide sequence of yeast centromere DNA was determined by Fitzgerald-Hayes et al. (1982). The centromere regions of two different yeast chromosomes were found to consist of an extremely A+T-rich core segment (87-88 bp in length) flanked by two identical short sequences. Bloom and Carbon (1982) demonstrated that the centromere DNA segment exhibiting the highest degree of sequence homology for the two yeast chromosomes was present in a unique and highly-ordered chromatin structure. This contained a discrete protected region of 220-250 bp. Topographical rearrangements of yeast centromere DNA segments were also investigated (Clarke and Carbon, 1983). Inversion of a centromere sequence or replacement with a sequence from another chromosome had no effect, thus indicating that yeast centromeres are not chromosome-specific.

The functioning of yeast and *Tetrahymena* telomere DNA sequences has been examined. The interest in the ends of linear DNA molecules stems from the intriguing problem of how these ends are replicated. Yeast and *Tetrahymena* telomeres were found to have a similar structure consisting of a simple repeated sequence containing nicks or gaps (Blackburn and Szostak, 1984). The addition of *Tetrahymena* telomeres to linear yeast plasmids did not affect the mitotic stability of the plasmids, which was believed to reflect the existence of spatial constraints upon the positions of telomeres for proper functioning (Dani and Zakian, 1983).

HISTONE-DEPLETED METAPHASE CHROMOSOMES AND CHROMOSOME SCAFFOLDS

Biochemical Properties

The complexity of metaphase chromosomes is the primary factor that hinders studies aimed at understanding their organization. How may this obstacle be surmounted? One approach has been to reduce the complexity of the problem by extracting proteins and thereby partly unfolding the chromosomes so that features of the arrangement of DNA fibers can be more easily observed. The residual proteins that maintain this higher level of DNA organization can also be readily identified. It is, of course, important that the extraction of proteins be carried out under mild conditions so that the resulting structures reflect the in situ arrangement of DNA. And since this procedure produces a major disruption of native chromosome morphology, care must be taken in interpreting the results to distinguish those aspects that reflect the in situ structure from those that may be artifactual. For this reason, it must be borne in mind that conclusions regarding DNA arrangement must be compatible with the results employing thin sectioning and scanning EM.

Applying the extraction procedure has provided extremely valuable information regarding the structural organization of metaphase chromosomes (Adolph et al., 1977a; Paulson and Laemmli, 1977). These studies utilized isolated HeLa metaphase chromosomes. Histones and most nonhistone proteins were removed at a pH near 7.0 with polyanions (dextran sulfate, heparin) or with high salt concentrations (2 M NaCl). The resulting histone-depleted chromosomes were characterized both biochemically, to determine which interactions maintain the residual organization, and structurally, to determine how the DNA is arranged. Histone-depleted HeLa chromosomes sedimented in sucrose density gradients as discrete peaks, demonstrating that the extraction procedure resulted in a homogeneous population of residual particles. The particles could be unfolded with proteases (chymotrypsin) but were not affected by ribonuclease. The proteins that were associated with histone-depleted chromosomes were characterized by SDS-polyacrylamide gel electrophoresis. Less than 0.1% of histones remained. No predominant nonhistones were found, but a number of minor species were present with molecular weights mainly above 50,000 daltons. These were designated "scaffolding proteins." The structure of the particles was examined in solution by fluorescence microscopy. Using a DNA stain of ethidium bromide, the extracted chromosomes were seen to consist of a halo of DNA surrounding a central region that retained the characteristic shape of the original chromosomes.

The presence of DNA is not essential to maintain the structural integrity of the residual protein scaffold (Adolph et al., 1977b). Digesting the DNA with nucleases such as micrococcal nuclease and DNase I before extracting proteins produced a structure that still possessed features of the original chromosomes. A fibrous network of residual nonhistones showed paired chromatids with the proportions of chromosomes, although expanded when examined by fluorescence microscopy or electron microscopy. The same subset of scaffolding nonhistones was found as for histone-depleted chromosomes.

The nature of the scaffolding proteins has been further examined by Lewis and Laemmli (1982), who showed that the scaffold is composed primarily of two high molecular weight species, Sc1 (170,000 daltons) and Sc2 (135,000 daltons). This study also demonstrated a role for metalloproteins, particularly Cu^{2+}-containing proteins, in stabilizing the protein structure. Earnshaw et al. (1984) have used autoimmune antisera against kinetochore proteins to reveal the presence of two of these proteins (of 77,000 daltons and 110,000 daltons) as components of the HeLa metaphase scaffold.

DNA Arrangement in Histone-Depleted Chromosomes

Besides identifying a subset of nonhistones responsible for maintaining the highest-order of chromosome structure, the other purpose of applying the histone-depletion procedure was to determine the arrangement of DNA in extracted chromosomes. Electron microscopy was required to resolve individual DNA strands in the halo surrounding the central scaffold (Paulson and Laemmli, 1977). Histone-depleted chromosomes were spread on an aqueous hypophase and prepared for electron microscopy by staining and rotary shadowing. For these particles, the halo of DNA was seen to extend an average of 10-12 μm radially from the central region. It was generally not possible to follow individual DNA strands for any distance because of the extensive overlap of fibers. However, exceptional micrographs were obtained that allowed the DNA to be traced as complete loops. For these chromosomes, most of the DNA was found to exist in loops at least 10-30 μm long, which corresponds to 30-90 kilobases. The number of average loop length was 14 μm (42 kb), while the average weighted with respect to the amount of DNA in a loop was 23 μm (70 kb). These values for DNA loop length should be considered minimum estimates. The same morphology for histone-depleted chromosomes resulted from using dextran sulfate/heparin or 2M NaCl to remove proteins. Little evidence of DNA supercoiling was apparent in the micrographs, perhaps because of nicking of the DNA, though a more recent EM study of histone-depleted chromosomes has revealed that the DNA is extensively supercoiled (Mullinger and Johnson, 1979).

The morphology of isolated scaffolds has been further characterized by electron microscopy (Earnshaw and Laemmli, 1983, 1984). The average chromatin loop length determined for intact chromosomes was $4.6 \pm 1.6 \mu$m, which corresponds to 83 ± 29 kb per loop. Residual scaffolds were seen to be fibrous structures that extend through the sister chromatids and that possess three distinct structural features. These were (a) peripheral material, (b) axial elements, and (c) apparent residual kinetochores. The scaffolds retained 5-10% of the total chromosome mass. Scaffolds could be selectively stained with silver, and thus appeared to be the target in intact chromosomes, which displayed an axial core by silver staining.

In Situ Existence of the Scaffold

Does the scaffold exist in situ, or is this structure artifactual, resulting, for example, from nonspecific aggregation of proteins (Okada and Comings, 1980; Hadlaczky et al., 1981)? The skepticism of some workers is partly based on the mistaken belief that the scaffold is proposed to be a dense, immutable, rod-like core of nonhistones that extends along the central axis of a chromosome. But it can be agreed that electron microscopy has demonstrated that metaphase chromosomes do not possess such a core. The scaffold, however, is not proposed to be a central core. Instead, scaffolding proteins should be considered to constitute a loose, fibrous network of linkers that extends through the entire chromosome. Some regions of the scaffold may be differentiated such that axial and peripheral elements may be present, but these proteins are not concentrated into a dense core. Indeed, with highly purified preparations, the scaffold is found to retain less than 5% of total chromosomal protein, and these proteins are distributed throughout the volume of a chromosome. Therefore, it is not surprising that visualizing the scaffold in intact chromosomes is extremely difficult.

Beyond this misconception concerning the appearance of the scaffold, evidence suggests that the scaffold is not artifactually created by nonspecific precipitation of chromosomal proteins. Histone-depleted chromosomes have a remarkably uniform structure, judging by their sedimentation behavior and appearance in the electron microscope, with a simple and reproducible composition of nonhistone proteins. The structures can be obtained from chromosomes isolated by a variety of procedures, which use aqueous buffers containing either divalent cations or polyamines, or which use hexylene glycol containing buffers. Histones and other proteins can be removed under low salt conditions with polyanions or removed by high salt concentrations (typically 2M NaCl). The fragile nature of the scaffold is, however, indicated by biochemical experiments showing a role for metalloproteins in stabilizing scaffold structure (Lewis and Laemmli, 1982).

The conclusion from investigating histone-depleted chromosomes that the DNA in metaphase chromosomes is primarily arranged in radial loops is supported by the EM studies of intact chromosomes. It is also compatible with the idea that DNA in the interphase nucleus is organized in looped domains, as described in a later section (and chap. 2). Indeed, the lengths of the looped domains that have been deduced for interphase DNA have similar values to the sizes of DNA loops in histone-depleted metaphase chromosomes, which range up to 90 kb.

Even if scaffolds were artifactually generated, which is contrary to the weight of evidence, characterizing histone-depleted metaphase chromosomes would still be extremely valuable. This is because these structures freeze the organization of DNA between the highly compact state of intact chromosomes and the indistinct tangle of DNA fibers of completely unfolded chromosomes. Thus, even if the scaffold interactions were entirely artifactual, the organization of DNA in histone-depleted chromosomes would still genuinely reflect the DNA packing in intact chromosomes. The observation of radial DNA loops as the predominant mode of DNA organization would maintain its validity and significance.

MODELS OF METAPHASE CHROMOSOME STRUCTURE

Radial Loop Model

Electron microscopy of intact chromosomes (Figs. 3-4 through 3-7) and histone-depleted chromosomes provides strong support for the concept that a fundamental mode of chromatin fiber organization in metaphase chromosomes is a radial array of loops. Thin sections of slightly swollen HeLa chromosomes cut transversely to the central chromatid axis clearly showed the radial orientation of fibers originating from close to the central axis (Fig. 3-4). Longitudinal sections demonstrated that this arrangement extends from telomere to telomere (Fig. 3-5). Scanning EM of isolated chromosomes revealed the existence of surface protuberances for slightly expanded chromosomes that seem to represent the peripheral tips of radial chromatin loops (Figs. 3-6, 3-7). The looped arrangement of chromosomal DNA was even more obvious with micrographs of histone-depleted chromosomes. For these extracted particles, the bulk of DNA was organized into loops of 30-90 kb extending from adjacent points along the central axis.

These observations suggest a radial loop model of metaphase chromosome structure, which is schematically illustrated in Figure 3-8. The predominant mode of chromatin packing is shown as loops, but, as Figure 8A indicates, longitudinal fibers could also exist near the central axis. These longitudinal fibers might serve to connect different segments along the chromatid axis. Since tracing chromatin fibers or DNA strands for any distance is difficult, it is not known how the loops are connected. In particular, it is not known whether any regularity or symmetry in the arrangement is present. Figure 8B illustrates a possible symmetric arrangement in which loops are organized into a shallow helix. The helix progresses from telomere to telomere along the direction of the chromatid axis. However, no evi-

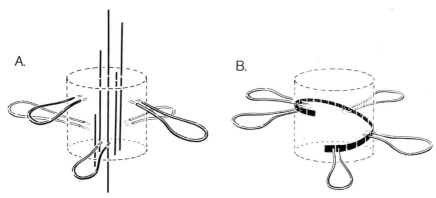

Figure 3-8. Schematic representation showing that the predominant mode of chromatin fiber organization in metaphase chromosomes is a radial array of loops. In (A), loops are indicated as the primary packing arrangement, though longitudinal fibers are also shown as possibly being present. The loops may have variable contour lengths, and the path of connecting the loops is unclear. In (B), the chromatin loops are shown as constituting a shallow helix. This hypothetical symmetric arrangement progresses along the direction of the chromatid axis.

dence has yet emerged for any regularity in the packing of fibers in metaphase chromosomes.

This organization of metaphase chromosome DNA into looped domains is compatible with the arrangement of DNA in interphase nuclei. Recent experiments have demonstrated that interphase DNA is not a diffuse tangle of strands, but is constrained into structural "domains" by the protein framework of the nucleus. The compartmentalization into domains creates interphase chromatin loops that are similar in size (approximately 80 kb) to their counterparts in metaphase.

The fundamental element of the "radial loop model" is that the predominant mode of organization of chromatin fibers is a looping arrangement that extends from close to the chromatid axis to the periphery. Another basic element of the model concerns the protein interactions, which have a role in establishing and maintaining this arrangement. A scaffold of nonhistones could maintain the structure and could be involved in the process of mitotic chromosome condensation, though this latter role is entirely speculative.

Helical Coil Models

The "radial loop model" shares many elements with the "folded-fiber model," in proposing that chromatin fibers are directly compacted into the characteristic metaphase chromosome morphology. No intermediate levels of structure between the 300 Å fibers and the assembled chromosomes are considered to exist. A radically different view of chromosome structure was presented by Bak et al. (1977) and Sedat and Manuelidis (1979). Central to this model is the presence of intermediate levels of helical coiling. Thus, Bak et al. (1977) suggested that the 300 Å nucleosomal helix is further coiled into a "supersolenoid" of 4000 Å diameter. A final coiling of the supersolenoid to produce a fivefold contraction of DNA length would produce the completed metaphase chromosome. This model has been further refined (Bak et al., 1979; Zeuthen et al., 1979). Sedat and Manuelidis (1978) also considered that metaphase chromosomes consist of a hierarchy of helices, but the dimensions of their proposed intermediate helices differed. According to their model, the nucleosomal helix of 300–500 Å is coiled into a tube having a diameter of 2000 Å, which is twisted once more to give the final chromosome width of 6000 Å. The model predicts that chromosomes have a hollow center with a diameter of about 2000 Å.

The experimental evidence for these helical coil models is primarily based on observations of (a) 4000 Å fibers in preparations of partially disintegrated chromosomes (Bak et al., 1977) and (b) dense tubes and spherical structures of 2000 Å diameter in interphase nuclei. However, both thin sections and scanning electron micrographs of intact HeLa metaphase chromosomes reveal little that suggests helical coiling (Figs. 3-4 through 3-7). G-banding techniques also show that contraction, and not coiling, is involved in the transition of chromosomes from prophase to metaphase (Yunis, 1976). The strongest evidence for coiling comes from light microscopy (Ohnuki, 1968). But these structures may result from the treatments, including acid fixation, that were used.

However, the "radial loop model" of chromosome structure can accommodate a level of helical coiling. Indeed, in Figure 3-8B the loops are indicated as originating from a helix progressing along the chromatid axis.

STRUCTURES OF OTHER CHROMOSOME TYPES

The concept that chromatin fibers in mitotic chromosomes are arranged as radial arrays of loops was largely developed from electron microscopic and biochemical studies of HeLa cell metaphase chromosomes. But the model gains substantial experimental support from observations of other chromosome types, and can also be applied to understand the structure of additional chromosomes. The "radial loop model" directly applies to other higher eukaryotic animal cells. The results of studies of chromosomes from Chinese hamster ovary cells, for example, were considered to be interchangeable with results for HeLa cells (e.g., Okada and Comings, 1979). An examination of the structure of mitotic chromosomes from mouse L cells and Novikoff rat hepatoma cells by thin sectioning and electron microscopy demonstrated the presence of radial distributions of fibers that appeared almost identical to the electron micrographs of HeLa chromosomes (Adolph, unpublished data).

The structure of lampbrush chromosomes of amphibian oocyte nuclei is analogous to the looped arrangement of mitotic chromosomes. The organization of lampbrush chromosomes can be observed more directly, however, since chromosomal fibers can be seen in the light microscope (Callan and Lloyd, 1960). These long-standing light microscope observations provide evidence for a looped arrangement of chromatin fibers, and provide a convincing parallel for a radial loop arrangement in HeLa and other mitotic chromosomes. This is because lampbrush chromosomes are found in a variety of animal and plant cells, and there is no evidence that these chromosomes represent an unusual type of structure (Callan and Lloyd, 1960). The mitotic arrangement would simply seem to be modified for the special requirements of early development in amphibian nuclei.

The meiotic chromosomes of *Bombyx mori* are a system in which structural transitions of loops can be traced leading to establishment of synaptonemal complexes (Rattner et al., 1980). Chromatin loops of early meiosis become progressively associated at their bases and form axial elements that are components of the synaptonemal complex. A central core is also observed in electron micrographs of mitotic chromosomes prepared from the milkweed bug *Oncopeltus fasciatus* (Foe et al., 1982). Loops of chromatin are organized by a central structure, which appears to stain distinctly because of the presence of a densely staining kinetochore.

RELATIONSHIP TO THE ARRANGEMENT OF DNA IN THE INTERPHASE NUCLEUS

What is the relationship between the extremely compact arrangement of chromatin in mitotic chromosomes and the relatively diffuse state of chromatin during interphase? Do chromosomes in interphase retain elements of organization that are related to the radial loop arrangement of mitotic chromosomes? Considerable evidence has recently been presented by a number of laboratories that interphase chromatin is compartmentalized into structural "domains." Nuclease digestion experiments, hydrodynamic measurements, and electron microscopy have shown that the domain size of interphase chromatin is from 75 to 220 kb, though values of around 85 kb are most common. This value is very similar to the average length of DNA loops in metaphase chromosomes (70–90 kb). It is beyond the scope of this review to extensively survey research concerned with the organization of DNA in the interphase nucleus and the role of the nuclear matrix. However, some evidence will be described concerning the relationship between DNA organization in interphase and metaphase, which comes from experiments in which nuclear proteins are extracted under mild conditions to produce "histone-depleted nuclei" and "nuclear scaffolds." These structures are the complement to "histone-depleted metaphase chromosomes" and "metaphase chromosome scaffolds" not only operationally, but also in their biochemical and structural properties.

Extracting histones and other nuclear proteins from isolated HeLa nuclei with polyanions (dextran sulfate/heparin) or high salt concentrations (2M NaCl) resulted in "histone-depleted nuclei." These particles still possessed a highly organized structure, as judged by sedimentation analysis and electron microscopy (Adolph, 1980). Histone-depleted nuclei sedimented homogeneously as discrete peaks in sucrose density gradients (at 12,000 S using 2M NaCl extraction and 6,000 S using polyanion extraction). These structures were unfolded by treatment with proteases (trypsin), while the sedimentation coefficient was reduced to a lesser degree with ribonuclease (RNase A). "Nuclear scaffolds" were produced by treating nuclei with DNase I or micrococcal nuclease before extracting proteins with 2M NaCl. Thin sectioning and electron microscopy showed, for DNase I and 2M NaCl extracted nuclei, the nuclear periphery and internal material to be present. Internal material was no longer observed when nuclei were treated with RNase A in addition to DNase I before protein extraction. SDS-polyacrylamide gel electrophoresis showed the major proteins of the "nuclear scaffold" to be the "lamins," the structural proteins of the peripheral nuclear lamina, and a number of minor, high molecular weight proteins.

Lebkowski and Laemmli (1982a, 1982b) were able to distinguish two types of histone-depleted nuclei. Type I structures had the biochemical and structural properties described previously. Type II structures were generated by including in the extraction buffer B-mercaptoethanol or dithiothreitol, or the metal chelators 1,10-phenanthroline or neocuproine. These particles sedimented at a lower rate than type I particles, apparently as a result of having a more expanded DNA halo. The presence of Cu or Ca converted type II particles to the type I form, suggesting that one level of DNA compaction in histone-depleted nuclei is stabilized by Cu or Ca. Nuclear lamina proteins and numerous high molecular weight species that comprised 10–15% of total nuclear proteins, were the major protein components of the type I structure. B-mercaptoethanol and 1,10-phenanthroline disrupted metalloprotein interactions of these particles to produce type II structures containing 3–5% of total nuclear proteins. Besides the lamins, two minor proteins of 64,000 and 200,000 daltons were present.

These results with histone-depleted nuclei and nuclear scaffolds clearly parallel the results with histone-depleted metaphase chromosomes and metaphase scaffolds. However, the observations cannot be directly compared since, for example, the lamins are not conserved as components of metaphase chromosomes and the metaphase proteins Sc1 and Sc2 are not recognizable among nuclear nonhistones. But the structural organization of DNA into looped domains is conserved from metaphase to interphase. The relationship between the factors responsible for the looped domain structure during the cell cycle remains to be unraveled.

REFERENCES

Adamietz, P.; Bredehorst, R.; and Hilz, H. (1978). ADP-ribosylated histone H1 from HeLa cultures. Fundamental differences to (ADP-ribose) $_n$-histone H1 conjugates formed *in vitro*. *Eur. J. Biochem.* **91,** 317-326.

Adolph, K. W. (1980a). Isolation and structural organization of human mitotic chromosomes. *Chromosoma (Berl.)* **76,** 23-33.

Adolph, K. W. (1980b). Organization of chromosomes in HeLa cells: Isolation of histone-depleted nuclei and nuclear scaffolds. *J. Cell Sci.* **42,** 291-304.

Adolph, K. W. (1981). A serial sectioning study of the structure of human mitotic chromosomes. *Eur. J. Cell Biol.* **24,** 146-153.

Adolph, K. W. (1984). Conservation of interphase chromatin nonhistone antigens as components of metaphase chromosomes. *FEBS Lett.* **165,** 211-215.

Adolph, K. W.; Cheng, S.M.; and Laemmli, U. K.; (1977a). Role of nonhistone proteins in metaphase chromosome structure. *Cell* **12**, 805-816.

Adolph, K. W.; Cheng, S. M.; and Paulson, J. R.; and Laemmli, U. K. (1977b). Isolation of a protein scaffold from mititic HeLa cell chromosomes. *Proc. Natl. Acad. Sci. (U.S.A.)* **74**, 4937-4941.

Adolph, K. W.; and Kreisman, L. R. (1983). Surface structure of isolated metaphase chromosomes. *Exp. Cell Res.* **147**, 155-166.

Adolph, K. W.; and Phelps, J. P. (1982). Role of non-histones in chromosome structure. Cell cycle variations in protein synthesis. *J. Biol. Chem.* **257**, 9086-9092.

Althaus, F. R.; Lawrence, S. D.; He, Y.-Z.; Stattler, G.L.; Tsukada, Y.; and Pitot, H. C. (1982). *Nature* **300**, 366-368.

Bahr, G. F. (1977). Chromosomes and chromatin structure. In *Molecular Structure of Human Chromosomes* (J. J. Yunis, ed.), Academic Press, New York.

Bak, A. L.; Zeuthen, J.; and Crick, F. H. C. (1977). Higher order structure of human mitotic chromosomes. *Proc. Natl. Acad. Sci. (U.S.A.)* **74**, 1595-1599.

Bak, P.; Bak, A. L.; and Zeuthen, J. (1979). Characterization of human chromosomal unit fibers. *Chromosoma (Berl.)* **73**, 301-315.

Blackburn, E. H.; and Szostak, J. W. (1984). The molecular structure of centromeres and telomeres. *Annu. Rev. Biochem.* **53**, 163-194.

Bloom, K. S.; and Carbon, J. (1982). Yeast centromere DNA is in a unique and highly ordered structure in chromosomes and small circular minichromosomes. *Cell* **29**, 305-317.

Blumenthal, A. B.; Dieden, J. D.; Kapp, L. N.; and Sedat, J. W. (1979). Rapid isolation of metaphase chromosomes containing high molecular weight DNA. *J. Cell Biol.* **81**, 255-258.

Cairns, J. (1966). Autoradiography of HeLa cell DNA. *J. Mol. Biol.* **15**, 372-373.

Callan, H. G.; and Lloyd, L. (1960). Lampbrush chromosomes of crested newts *Triturus cristatus (Laurenti). Philos. Trans. R. Soc. Lond.* **B243**, 135-219.

Caspersson, T.; Hulten, M.; Lindsten, J.; and Zech, L. (1971). Identification of chromosome bivalents in human male meiosis by quinacrine mustard fluorescence analysis. *Hereditas* **67**, 147-149.

Cervenka, J.; Thorn, H. L.; and Gorlin, R. J. (1973). Structural basis of banding pattern of human chromosomes. *Cytogenet. Cell Genet.* **12**, 81-86.

Christenhuss, R.; Buchner, T.; and Pfeiffer, R. A. (1967). Visualization of human somatic chromosomes by scanning electron microscopy. *Nature* **216**, 379-380.

Clarke, L.; and Carbon, J. (1983). Genomic substitutions of centromeres in *Saccharomyces cerevisiae. Nature* **305**, 23-28.

Compton, J. J.; Hancock, R.; Oudet, P.; and Chambon, P. (1976). Biochemical and electron microscopic evidence that the subunit structure of Chinese-hamster-ovary interphase chromatin is conserved in mitotic chromosomes. *Eur. J. Biochem.* **70**, 555-568.

Cook, P. J. L.; and Hamerton, J. L. (1979). Report of the committee on the genetic constitution of chromosome 1. *Cytogenet. Cell Genet.* **25**, 9-20.

Dani, G.; and Zakian, V. A. (1983). Mitotic and meiotic stability of linear plasmids in yeast. *Proc. Natl. Acad. Sci. (U.S.A.)* **80**, 3406-3410.

Daskal, Y.; Mace, M. L.; Wray, W.; and Busch, H. (1976). Use of direct current sputtering for improved visualization of chromosome topography by scanning electron microscopy. *Exp. Cell Res.* **100**, 204-212.

DuPraw, E. J. (1966). Evidence for a "folded-fiber" organization in human chromosomes. *Nature* **209**, 577-581.

DuPraw, E. J. (1970). *DNA and Chromosomes.* Holt, Rinehart and Winston, New York.

Earnshaw, W. C.; and Laemmli, U. K. (1983). Architecture of metaphase chromosomes and chromosome scaffolds. *J. Cell Biol.* **96**, 84-93.

Earnshaw, W. C.; and Laemmli, U. K. (1984). Silver staining the chromosome scaffold. *Chromosoma (Berl.)* **89,** 186-192.
Earnshaw, W. C.; Halligan, N.; Cooke, C.; and Rothfield, N. (1984). The kinetochore is part of the metaphase chromosome scaffold. *J. Cell Biol.* **98,** 352-357.
Farzaneh, F.; Zalin, R.; Brill, D.; and Shall, S. (1982). DNA strand breaks and ADP-ribosyl transferase activation during cell differentiation. *Nature* **300,** 362-366.
Fitzgerald-Hayes, M.; Clarke, L.; and Carbon, J. (1982). Nucleotide sequence comparisons and functional analysis of yeast centromere DNAs. *Cell* **29,** 235-244.
Foe, V. E.; Forrest, H.; Wilkinson, L.; and Laird, C. (1982). In *Insect Ultrastructure* (H. Akai and R. King, eds.), vol. 1, pp. 222-246, Plenum, New York.
Gall, J. G. (1963). Kinetics of deoxyribonuclease action on chromosomes. *Nature* **198,** 36-38.
Gall, J. G. (1966). Chromosome fibers studied by a spreading technique. *Chromosoma (Berl.)* **20,** 221-233.
Golomb, H. M. (1976). Human chromatin and chromosomes studied by scanning electron microscopy: Progress and perspectives. *J. Reprod. Med.* **17,** 29-35.
Golomb, H. M.; and Bahr, G. F. (1971). Scanning electron microscopic observations of surface structure of isolated human chromosomes. *Science* **171,** 11024-11026.
Hadlaczky, G.; Sumner, A. T.; and Ross, A. (1981). Protein-depleted chromosomes II. Experiments concerning the reality of chromosome scaffolds, *Chromosoma (Berl.)* **81,** 557-567.
Harrison, C. J.; Britch, M.; Allen, T. D.; and Harris, R. (1981). Scanning electron microscopy of the G-banded human karyotype. *Exp. Cell Res.* **134,** 141-153.
Harrison, C. J.; Allen, T. D.; Britch, M.; and Harris, R. (1982). High-resolution scanning electron microscopy of human metaphase chromosomes. *J. Cell Sci.* **56,** 409-422.
Harrison, C. J.; Allen, T. D.; and Harris, R. (1983). Scanning electron microscopy of variations in human metaphase chromosome structure revealed by Giemsa banding. *Cytogenet. Cell Genet.* **35,** 21-27.
Hayaishi, O.; and Ueda, K. (1977). Poly(ADP-ribose) and ADP-ribosylation of proteins. *Annu. Rev. Biochem.* **46,** 95-116.
Hilz, H.; and Stone, P. (1976). Poly (ADP-ribose) and ADP-ribosylation of proteins. *Rev. Physiol. Biochem. Pharmacol.* **76,** 1-58.
Holtlund, J.; Kristensen, T.; Ostvold, A. C.; and Laland, S. G. (1983). ADP-ribosylation in permeable HeLa S3 cells. *Eur. J. Biochem.* **130,** 47-51.
Huberman, J. A.; and Riggs, A. D. (1966). Autoradiography of chromosomal DNA fibers from Chinese hamster cells. *Proc. Natl. Acad. Sci. (U.S.A.)* **55,** 599-606.
Johnstone, A. P.; and Williams, G. T. (1982). Role of DNA breaks and ADP-ribosyl transferase activity in eukaryotic differentiation demonstrated in human lymphocytes. *Nature* **300,** 368-370.
Jump, D. B.; and Smulson, M. (1980). Purification and characterization of the major nonhistone protein acceptor of poly(adenosine diphosphate ribose) in HeLa cell nuclei. *Biochemistry* **19,** 1024-1030.
Jump, D. B.; Butt, T. R.; and Smulson, M. (1979). Nuclear protein modification and chromatin substructure. 3. Relationship between Poly(adenosine diphosphate) ribosylation and different functional forms of chromatin. *Biochemistry* **18,** 983-990.
Kavenoff, R.; and Zimm, B. H. (1973). Chromosome-sized DNA molecules from *Drosophila*. *Chromosoma (Berl.)* **41,** 1-27.
Kavenoff, R.; Klotz, L. C.; and Zimm, B. H. (1974). On the nature of chromosome-sized DNA molecules. *Cold Spring Harbor Symp. Quant. Biol.* **38,** 1-8.
Korp, B. R.; and Diacumakos, E. G. (1978). Microsurgically-extracted metaphase chromosomes of the Indian Muntjac examined with phase contrast and electron microscopy. *Exp. Cell Res.* **111,** 83-93.

Langmore, J. P.; and Paulson, J. R. (1983). Low angle X-ray diffraction studies of chromatin structure *in vivo* and in isolated nuclei and metaphase chromosomes. *J. Cell Biol.* **96**, 1120-1131.

Lebkowski, J. S.; and Laemmli, U. K. (1982a). Evidence for two levels of DNA folding in histone-depleted HeLa interphase nuclei. *J. Mol. Biol.* **156**, 309-324.

Lebkowski, J. S.; and Laemmli, U. K. (1982b). Non-histone proteins and long-range organization of HeLa interphase DNA. *J. Mol. Biol.* **1569**, 325-344.

Lewis, C. D.; and Laemmli, U. K. (1982). Higher order metaphase chromosome structure: Evidence for metalloprotein interactions. *Cell* **29**, 171-181.

Marsden, M. P. F.; and Laemmli, U. K. (1979). Metaphase chromosome: Evidence for a radical loop model, *Cell* **17**, 849-858.

Mullinger, A. M.; and Johnson, R. T. (1979). The organization of supercoiled DNA from human chromosomes. *J. Cell Sci.* **38**, 369-389.

Ohnuki, Y. (1968). Structure of chromosomes. I. Morphological studies of the spiral structure of human somatic chromosomes. *Chromosoma (Berl.)* **25**, 402-428.

Okada, T. A.; and Comings, D. E. (1979). Higher order structure of chromosomes. *Chromosoma (Berl.)* **72**, 1-14.

Okada, T. A.; and Comings, D. E. (1980). A search for protein cores in chromosomes: Is the scaffold an artefact? *Am. J. Hum. Genet.* **32**, 814-832.

Paulson, J. R.; and Laemmli, U. K. (1977). The structure of histone-depleted metaphase chromosomes. *Cell* **12**, 817-828.

Paulson, J. R.; and Langmore, J. P. (1983). Low angle X-ray diffraction studies of HeLa metaphase chromosomes: Effects of histone phosphorylation and chromosome isolation procedure. *J. Cell Biol.* **96**, 1132-1137.

Prescott, D. M. (1970). The structure and replication of eukaryotic chromosomes. In *Advances in Cell Biology* (D. M. Prescott, L. Goldstein, and E. McConkey, eds.), vol. 1, pp. 57-117, North Holland, Amsterdam.

Purnell, M. R.; Stone, P. R.; and Whish, W. J. D. (1980). ADP-ribosylation of nuclear proteins. *Biochem. Soc. Trans.* **8**, 215-227.

Rattner, J. B.; Goldsmith, M.; and Hamkalo, B. A. (1980). Chromatin organization during meiotic prophase of *Bombyx mori*. *Chromosoma (Berl.)* **79**, 215-224.

Ris, H.; and Witt, P. L. (1981). Structure of the mammalian kinetochore. *Chromosoma (Berl.)* **82**, 153-170.

Scheid, W.; and Traut, H. (1971). Visualization by scanning electron microscopy of achromatic lesions ("gaps") induced by X-rays in chromosomes of *Vicia faba*. *Mutation Res.* **11**, 253-255.

Sedat, J.; and Manuelidis, L. (1978). A direct approach to the structure of eukaryotic chromosomes. *Cold Spring Harbor Symp. Quant. Biol.* **42**, 331-250.

Song, M.-K. H.; and Adolph, K. W. (1983a). ADP-ribosylation of nonhistone proteins during the HeLa cell cycle. *Biochem. Biophys. Res. Commun.* **115**, 938-945.

Song, M.-K. H.; and Adolph, K. W. (1983b). Phosphorylation of nonhistone proteins during the HeLa cell cycle. Relationship to DNA synthesis and mitotic chromosome condensation. *J. Biol. Chem.* **258**, 3309-3318.

Stubblefield, E.; and Wray, W. (1971). Architecture of the Chinese hamster metaphase chromosome. *Chromosoma (Berl.)* **32**, 262-294.

Tanuma, S.-I.; and Kanai, Y. (1982). Poly(ADP-ribosyl)ation of chromosomal proteins in the HeLa S3 cell cycle. *J. Biol. Chem.* **257**, 6565-6570.

Taylor, J. H.; Woods, P. S.; and Hughes, W. L. (1957). The organization and duplication of chromosomes as revealed by autoradiographic studies using tritium-labeled thymidine, *Proc. Natl. Acad. Sci. (U.S.A.)* **43**, 122-128.

Waldren, C. A.; and Johnson, R. T. (1974). Analysis of interphase chromosome damage by means of premature chromosome condensation after X- and ultraviolet-irradiation. *Proc. Natl. Acad. Sci. (U.S.A.)* **71,** 1137-1141.

Wong, M.; Kani, Y.; Miwa, M.; Bustin, M.; and Smulson, M. (1983). Immunological evidence for the *in vivo* occurrence of a crosslinked complex of poly (ADP-ribosylated) histone H1. *Proc. Natl. Acad. Sci. (U.S.A.)* **80,** 205-209.

Wray, W.; and Stubblefield, E. (1970). A new method for the rapid isolation of chromosomes, mitotic apparatus or nuclei from mammalian fibroblasts at near neutral pH. *Exp. Cell Res.* **59,** 469-478.

Wray, W.; Stubblefield, E.; and Humphrey, R. (1972). Mammalian metaphase chromosomes with high molecular weight DNA isolated at pH 10.5. *Nature* **238,** 237-238.

Yunis, J. J. (1976). High resolution of human chromosomes. *Science* **191,** 1268-1270.

Zeuthen, J.; Bak, P.; and Bak, A. L. (1979). Chromosomal unit fibers in *Drosophila. Chromosoma (Berl.)* **73,** 317-326.

4

The Organization of Meiotic Chromosomes and Synaptonemal Complexes

Michael S. Risley
*Cornell University Medical College
New York, New York*

During the life cycle of sexually reproducing organisms, haploid gametes containing new combinations of the parental genomic information are generated by the process of meiosis. Chromosome interactions occur during meiosis in a specialized sequence that results in the pairing and exchange of information between homologues, and permits the disjunction of the homologues during division so that the haploid products are balanced chromosomally. Recombination and disjunction both appear to be dependent upon chromosome pairing and the maintenance of the paired state, since asynaptic or desynaptic mutants exhibit reduced cross-over frequencies and elevated levels of nondisjunction (Baker et al., 1976; Golubovskaya, 1979). Aspects of chromosome organization and behavior that are important for pairing and information exchange are therefore critical to meiosis.

MEIOSIS: AN OVERVIEW

Meiosis occurs in all sexually reproducing organisms and, although there are certain details that are species specific, it is highly conserved in evolution. The process begins following a special period of DNA replication (premeiotic S) in the stem cells. Premeiotic S is followed by a prolonged (relative to mitosis) meiotic prophase, and meiosis is then completed by two successive cell divisions that lack an intervening S phase.

Events that occur during premeiotic S appear to play important roles in meiosis, since treatment of premeiotic S cells with metabolic inhibitors, colchicine or radiation can impair crossing over and chiasma formation (see Henderson, 1970; Stern et al., 1975). One explanation for this may be that chromosome pairing and recombination occur during premeiotic S, and that these processes are directly affected by the treatments. Evidence for this view includes the observation that recombination deficient, temperature sensitive mutants of *Drosophila melanogaster* females are most sensitive to temperature shifts during the pro-oocyte, premeiotic S phase (Grell, 1978; Grell and Generoso, 1980). Electron microscopy of the developing pupal germarium in *Drosophila melanogaster* resulted in the identification of synaptonemal complexes in the pro-oocytes (Grell and Generoso, 1982). This was interpreted as evidence for premeiotic synapsis. Synaptonemal complexes have also been reported to be present in premeiotic S cells from wheat (McQuade and Pickles, 1980) and male mice (Grell et al., 1980).

Conventional schemes for meiosis place the time of synapsis and recombination in meiotic prophase, not premeiotic S. This is due to the observation that synaptonemal complexes are usually first assembled during meiotic prophase (Moses, 1968; Westergaard and von Wettstein, 1972; von Wettstein et al., 1984). Three dimensional reconstructions from serial sections of leptotene nuclei (which lack synaptonemal complexes) have failed to find evidence for paired chromosomes prior to prophase (Rasmussen and Holm, 1980). There is also some controversy regarding the accuracy of identification of premeiotic S phase cells (Bennett et al., 1979a; Carpenter, 1981). It is conceivable that synapsis or recombination could begin during premeiotic S in some organisms, particularly in insects that have extensively paired chromosomes in somatic cells. This does not appear to be generally true, however, and it is clearly not true in haploid organisms that replicate the homologues in separate nuclei preceeding karyogamy (Carmi et al., 1978). Molecular events in premeiotic S of most organisms appear to predispose chromosomes to synapsis and recombination during the subsequent meiotic prophase. The nature of these important premeiotic S processes is not understood, but they may be associated with special characteristics of DNA replication during this phase.

DNA replication during premeiotic S is unique in several ways. First, the synthetic phase is usually 3-6 times longer than mitotic S. Premeiotic S lasts 14 hours in male mice (Monesi, 1962), 50 hours in *Lilium* microsporocytes (Holm, 1977a), and 9-10 days in male *Triturus* (Callan and Taylor, 1968). The prolonged replication is primarily a consequence of a relatively low rate of initiation of DNA replication during premeiotic S (Callan, 1972). It may also be lengthened, as in *Lilium*, by the occurrence of a nonsynthetic gap during the S phase (Holm, 1977a). Another unique characteristic of premeiotic S, at least in the plants *Lilium* and *Trillium* (Hotta and Stern, 1971; Ito and Hotta,

1973), is that DNA is not completely replicated. In the lily, 0.2-0.3% of the DNA remains unreplicated until meiotic prophase, when replication is completed at zygotene.

Meiotic prophase begins after a brief G2 phase. According to cytological criteria (chromosome morphology and phase of chromosome synapsis), prophase is divided into several substages (see Fig. 4-1). The first stage, leptotene, is characterized by the appearance of long, slender chromosomes, attached at their telomeres to the nuclear membrane, and at numerous points along their length to an axial protein core. Early in leptotene, the telomeres are randomly dispersed around the inner nuclear membrane, and homologous telomeres are not associated (Moens, 1969a; Holm, 1977b; Rasmussen and Holm, 1978a). In many organisms the telomeres move over the surface of the inner nuclear membrane during leptotene, and concentrate in one region, forming the classical bouquet configuration. The homologous telomeres are thereby brought closer together, presumably to promote alignment. Formation of the bouquet is not essential for meiosis, since there are species that do not form bouquets—for example, *Coprinus* (Holm et al., 1981), and *Neurospora* (Gillies, 1979).

When homologous regions are brought into close alignment (300 nm), a lattice of filaments assembles on the aligned protein cores and connects them, forming ladderlike structures known as synaptonemal complexes (SCs). This marks the beginning of chromosome synapsis and the zygotene

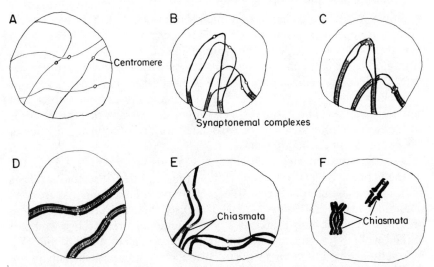

Figure 4-1. Schematic overview of chromosome mechanics during meiotic prophase. A, leptotene; B, early zygotene; C, late zygotene; D, pachytene; E, diplotene; F, diakinesis.

stage of meiosis, both of which are completed when SCs have been assembled along the entire length of each bivalent. The process of synapsis is DNA sequence specific during zygotene, and commonly begins preferentially in telomeric regions and subsequently spreads to other regions of the homologues as these regions become closely aligned (e.g., *Bombyx*: Holm and Rasmussen, 1980; Human spermatocytes: Rasmussen and Holm, 1978*a*). In some organisms (e.g., *Lilium*; see Holm, 1977*a*), however, synapsis is initiated at multiple sites of alignment along the forming bivalent.

Since individual chromosomes are not paired closely prior to SC formation, chromosomes and bivalents can become intertwined or interlocked during synapsis. This is particularly evident in zygotene nuclei from organisms with long chromosomes (reviewed in Rasmussen and Holm, 1980; von Wettstein et al., 1984). Resolution of interlocking occurs during zygotene, thereby preventing the formation of multivalents at later meiotic stages. The DNA and lateral elements of interlocked chromosomes break to permit passage of one chromosome or bivalent past the other (Holm, 1977*b*; Rasmussen and Holm, 1978*a*). The broken ends then appear to realign and rejoin. Resolution may also occur in some instances by the release of telomeres from the nuclear membrane, permitting chromosome movement and disentanglement (Rasmussen and Holm, 1980).

Pachytene follows zygotene, and is generally considered to be the stage when crossing over occurs and chiasma form at the cross-over sites. The synapsis of homologous regions is complete by pachytene, but synapsis of inter- or intra-chromosomal nonhomologous DNA sequences may take place. This nonspecific pairing has been observed in haploid plant meiocytes (Gillies, 1974), in triploid oocytes from the silk worm, *Bombyx mori* (Rasmussen, 1977*b*), and during the "synaptic adjustment" of tandem duplications in mouse spermatocytes (Moses and Poorman, 1981). SCs assemble in the paired regions containing nonhomologous sequences and, therefore, are not the determinants of specific pairing. Pachytene is also the stage when the chromosomal bouquet is lost as a consequence of the movement of the telomeres to more random positions around the nuclear membrane.

Diplotene is the next prophase stage, and is characterized by the disassembly of the SCs, except where crossing over has occurred and chiasma have formed. As desynapsis occurs, the homologues move away from each other, but remain attached by the chiasmata. During the next prophase stage, diakinesis, the chromosome arms rotate, the chiasmata terminalize (in some organisms), the nuclear membrane disassembles, and the chromosomes associate with the spindle apparatus. The centromeres of each homologue in a bivalent align on the spindle so that they face opposite poles during metaphase I. At anaphase, the homologues disjoin and are segregated to opposite poles. The sister chromatids separate only during the second meiotic division, which is similar to a mitotic division since the centromeres split at this time.

The special organization and behavior of meiotic chromosomes permits the formation of stable (several days to two weeks) pairing relationships between homologues. Certain aspects of chromosome pairing during meiosis may be related to mitotic chromosome pairing or "somatic association," which can be seen cytologically in *Dipteran* or plant cells (reviewed in Comings, 1980). The relatively extensive chromosome pairing and recombination in meiotic cells, however, are dependent upon the formation of structures specific to meiosis, the SCs.

SYNAPTONEMAL COMPLEXES

Structure and Assembly

The view that the SCs play a fundamental role in meiosis is partly based on the widespread occurrence of these organelles in both the plant and animal kingdoms. This view is reinforced by the observation that meiotic chromosomes in spermatocytes of *Drosophila melanogaster* lack synaptonemal complexes and fail to cross over (Rasmussen, 1973). Moreover, the aberrant assembly of the SCs that frequently occurs in meiotic mutants, such as the asynaptic (as) mutants in plants (Swaminathan and Murty, 1959; Martini and Bozzini, 1966; Stringham, 1970), or the c3G (Smith and King, 1968; King, 1970; Rasmussen, 1975a) and ord (Lindsley and Sandler, 1977) mutants in *Drosophila* oocytes, is associated with meiotic impairment and structural alterations of chromosomes (reviewed in Baker, 1976; Golubovskaya, 1979). SC assembly appears to be essential to achieve stable synapsis and crossing over; however, SC formation is not sufficient to achieve crossing over, since SCs form in *Bombyx* oocytes between homologues that do not cross over (Rasmussen, 1976, 1977b), and in *Drosophila* oocytes with mutations that significantly reduce crossing over and chiasma formation (Carpenter, 1979).

SCs were first described by Moses (1956) and Fawcett (1956) in ultrastructural studies of spermatocyte nuclei in which conventional transmission electron microscopy was employed. Further development of knowledge regarding SC structure and assembly was greatly facilitated by the use of two different microscopic approaches. In the first, testes or ovaries are serially sectioned, and the sections examined and photographed with a transmission electron microscope. The photographs are then traced, and three dimensional reconstructions of meiotic nuclei are prepared (e.g., Wettstein and Sotelo, 1967; Moens, 1969a; Gillies, 1972; Rasmussen, 1976). This procedure, although tedious, has yielded substantial insights into SC ultrastructure and the sequence of SC assembly, chromosome pairing, and chiasma formation.

The second major approach to SC analysis has been to swell and spread meiotic cells on a liquid-air interphase, attach them to coated slides and grids, and examine them microscopically after staining with ethanolic phosphotunstic acid (Counce and Meyer, 1973) or silver (Pathak and Hsu, 1979; Fletcher, 1979). These microspreading methods can be used to study the same nucleus by light and electron microscopy if silver staining is employed (Dresser and Moses, 1979, 1980). This method is also relatively rapid compared to 3-D reconstruction and, therefore, useful for karyotype analysis (Moses, 1977a; Dresser and Moses, 1979).

When completely assembled at pachytene, each SC resembles a ladderlike ribbon about 0.2 μm wide. They extend the full length of each bivalent (except the XY pair in mammals), and usually exhibit several twists in each bivalent (see Fig. 4-2). Although the morphological details of the SCs vary somewhat between species (see Moses, 1968; Westergaard and von Wettstein, 1972), two major structural components of SCs can usually be discerned: two lateral elements, one from each homologue, and the central components.

The lateral elements in an SC are each 300-500 Å wide, and are separated by a gap of 100-150 nm. Each homologue is attached to a separate lateral element, and the lateral elements are attached at their ends to the inner nuclear membrane via a plaque of filamentous material or terminal swellings (Wettstein and Sotelo, 1967; Woolam et al., 1967; Moens, 1969a; Esponda and Giménez-Martin, 1972). In sections, the lateral elements appear as rods comprised of interwound fibrous subunits that often display a discrete structural periodicity (Moses and Coleman, 1964; Moens, 1969a; Westergaard and von Wettstein, 1970). Using photographic superposition and enhancement techniques, Stubblefield (1973) has shown that major structural repeats

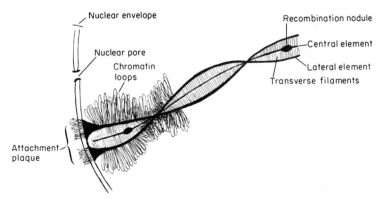

Figure 4-2. Schematic representation of the organization of a synaptonemal complex. Note that only a portion of the chromatin loops has been presented; in situ, the chromatin loops radiate from each lateral element in all directions except towards the central region.

occur at 85 and 340 Å in *Locusta*, 340 Å in rat, and 370 Å in *Neotiella*. The repeat patterns were seen in both unpaired and paired lateral elements, indicating that the repeat distance reflected the ordered assembly of lateral element macromolecules.

Occasionally sectioned material (see Moses, 1968) and, more often, whole mounts of spread pachytene nuclei (Comings and Okada, 1970; Counce and Meyer, 1973; Tres, 1977; Wahrman, 1981) reveal each lateral element to be comprised of two longitudinal subfibers, particularly in telomeric regions. The double nature of each lateral element has been accounted for by assuming the presence of a separate axial element for each sister chromatid in a homologue (Comings and Okada, 1971; Tres, 1977; Wahrman, 1981). However, the doubleness is not usually seen during leptotene and early zygotene. Moreover, lateral elements in late pachytene bivalents often appear to consist of three or more subfibers (Moses et al., 1975; Solari, 1980).

Lateral elements begin assembling during leptotene as the axial cores of chromosomes. The axial cores assemble medially along the longitudinal axis of each chromosome. The chromatin then rotates around the longitudinal axis during late leptotene and displaces the core to a lateral position, where it is exposed to facilitate binding to the SC central components. In some species (e.g., *Lilium*; see Holm, 1977a), the axial cores or lateral element are completely assembled prior to zygotene. In other species (e.g., *Bombyx mori*: Holm and Rasmussen, 1980; humans: Rasmussen and Holm, 1978a; Bojko, 1983), the lateral elements remain incomplete until sometime during zygotene.

Central components of the SC lie medially between the two lateral elements. Central components include thin (16-18 Å) transverse filaments (Solari and Moses, 1973) oriented perpendicular to the lateral elements, and a central element that varies morphologically between species and ranges in thickness from 150 to 500 Å. The central element may appear as a single, longitudinally-oriented filament, or as a complex tripartite structure consisting of a lattice of interconnected longitudinally-oriented filaments (e.g., *Neotiella*: Westergaard and von Wettstein, 1970; *Bombyx*: King and Akai, 1971). The central components are connected to each lateral element by the thin filaments.

In addition to the major structural components just described, there are other components or structural modifications associated with the SCs. One of these if the kinetochore. In spread whole mounted meiocytes stained with ethanolic phosphotungstic acid, residual kinetochores can be visualized bound to the lateral elements of the SC (Counce and Meyer, 1973). The nature of the SC-kinetochore interaction is not known, but the presence of the kinetochore facilitates karyotype analysis, since arm ratios can be measured when centromere position is known. Structural modifications of the lateral elements, in the form of swellings or bulges, have also been noted in lily (Moens, 1970), *Sordaria* (Zickler, 1973; Zickler and Sage, 1981), maize (Gillies, 1973) and especially in the sex chromosomes of several mammals (Solari, 1974; Moses, 1977b). The significance of these structures is unknown.

Recombination nodules are structures associated with the central components of the SC, and appear to be important, as the name implies, in recombination and chiasma formation. They are present on the SCs during pachytene in all recombination proficient organisms studied from both the plant and animal kingdoms. The nodules, first observed in *Drosophila* oocytes by Carpenter (1975), vary in morphology within and between species, but are usually round (30-70 Å diameter) or ovoid (20-90 Å wide, 30-200 Å long). The time of their appearance during prophase also varies between species. In *Ascomycetes* (Zickler, 1977; Gillies, 1979) and *Bombyx* spermatocytes (Holm and Rasmussen, 1980), they are present from zygotene through diplotene. In *Ascaris* (Kundu and Bogdanov, 1979), they are present only from pachytene through diplotene, while human spermatocytes have nodules only at zygotene and pachytene (Rasmussen and Holm, 1978a).

The principal evidence for a role of the nodules in recombination has been the correlation between the number and location (which is nonrandom) of nodules with the frequency and sites of crossing over and chiasma formation (for review see Carpenter, 1979; von Wettstein et al., 1984). Nodules have been found to be absent in the recombination deficient oocytes of *Bombyx mori*, which have normal SCs (Rasmussen, 1976). Carpenter (1979) has also shown that there is a close correspondence between the reduced cross-over frequencies and nodule frequencies in the meiotic mutants mei-41 and mei-218 of *Drosophila* oocytes. Electronmicroscopic autoradiography has identified recombination nodules in *Drosophila* oocytes as sites of DNA synthesis, a result expected if the nodules contain the enzymatic machinery necessary for DNA recombination (Carpenter, 1981). The role of the nodules in recombination is unclear, partly because of their variability in size, shape, and number during meiosis. Further investigation of the appearance and fate of nodules during meiosis in an organism with synchronously developing meiocytes may provide important insights to nodule function.

Since the SC has not been characterized biochemically, little is known of the mechanisms of assembly and disassembly. The lateral elements appear during leptotene or early zygotene; therefore, the macromolecules in these structures must be synthesized prior to or during these stages. Macromolecules of the central components must also be synthesized prior to or during zygotene. A pre-zygotene synthesis of central components is suggested by the observation that material morphologically similar to the central components of the SC assemble into extrachromosomal structures prior to zygotene in yeast (Moens and Rappaport, 1971), *Neotiella* (Westergaard and von

Wettstein, 1970), and *Ascaris* (Bogdanov, 1977; Fiil, 1978). Moreover, transverse filaments are associated with leptotene chromosomes in lily meiocytes (Moens, 1968). The extrachromosomal complexes often assemble in the nucleoli in yeast and *Neotiella*, and in the cytoplasm near the nuclear membrane, in *Ascaris*. As the SCs assemble during zygotene, the extrachromosomal complexes disappear, suggesting a precursor-product relationship. Synthesis of SC components may also occur during pachytene to provide a supply of precursors for SC assembly during synapsis of nonhomologous chromosome regions.

Disassembly of the SCs occurs during diplotene. SC central components dissociate from the bivalents in all regions except where crossing over has occurred and chiasmata have formed (Westergaard and von Wettstein, 1970; Zickler, 1977; Rasmussen and Holm, 1978b; Gillies, 1979). These residual SC fragments often have associated recombination nodules, and may remain bound to the chromosomes through metaphase I, maintaining the alignment of the homologues until disjunction occurs (Rasmussen and Holm, 1978b; Moens and Church, 1979).

The disassembly of the SC is frequently accompanied by the appearance, in the nucleus or cytoplasm, of extrachromosomal structures termed polycomplexes. They have been observed most often in insect meiocytes (Roth, 1966; Moens, 1969b; Fiil and Moens, 1973; Sotelo et al., 1973; Rasmussen, 1975b; Esponda and Krimer, 1979), but are also present in meiocytes of plants (Moens, 1968; Kehlhoffner and Dietrich, 1983) and in human spermatocytes (Solari and Vilar, 1978). Polycomplexes have been shown to persist into postmeiotic stages in certain species (Moens, 1969b; Solari and Vilar, 1978; Esponda and Krimer, 1979). The complexes appear to be comprised of multiple SC components that have been shed from the diplotene bivalents. It is not clear, however, if the structures derive from SCs that were shed intact, from SC subunits that reassemble after they dissociate from the chromosomes, or from excess subunits that were stored in the meiocytes or synthesized de novo at diplotene.

Chemistry

Cytochemical studies suggest that the SC consists primarily of protein. The structural integrity of the SC is dependent upon proteins, since proteases destroy SCs while DNase and RNase appear to have little effect on SC structure (Comings and Okada, 1970; Coleman and Moses, 1964; Solari, 1972; Comings and Okada, 1976). The protein interactions are probably stabilized by hydrophobic or hydrogen bonds since the SC is stable in 2 M NaCl (Solari, 1972; Comings and Okada, 1976), but solubilized in 8 M urea (Comings and Okada, 1970).

Central components are less stable than the lateral elements. Solutions containing less than 0.45% NaCl destabilize the central components, while the lateral elements are resistant to distilled water, as well as low concentrations of Joy detergent (Weith and Traut, 1980) or SDS (Ierardi et al., 1983). Lateral elements appear to be enriched for basic proteins, since they stain preferentially with silver (Westergaard and von Wettstein, 1970; Dresser and Moses, 1979, 1980) and ethanolic phosphotunstic acid (Sheridan and Barnett, 1969; Counce and Meyer, 1973). Solutions of dilute HCL will also destroy the SC (Sheridan and Barnett, 1969; Comings and Okada, 1979).

The structural proteins of the SC have not been identified. Currently, several laboratories are attempting to identify SC proteins by immunochemical methods (De Martino et al., 1980; Dresser and Moses, 1983; Moses, 1985), and by isolating SCs for direct biochemical characterization (Gambino et al., 1981; Ierardi et al., 1983; Li et al., 1983). De Martino et al. (1980) used immunofluorescence microscopy to study the binding of antibodies to myosin and actin in spermatocytes of the rat. Antimyosin stained the central region of the pachytene bivalents, while antiactin appeared to stain the telomere-membrane attachment sites. Since ultrastructural localization was not conducted in these studies, the relationship between myosin, actin, and the SCs remains unclear. Dresser and Moses (1983) have reported the isolation of a monoclonal antibody (SC3) from autoimmune mice that binds to the central components of SCs. The antibody also bound to structures present in spermatids and to intermediate filaments in the cytoplasm of fibroblasts (Moses, 1985).

The dimensions of the total SC complement in pachytene nuclei (e.g., 25 μm in yeast, 236 μm in humans, 3700 μm in lilies: from von Wettstein et al., 1984) suggest that some of the proteins of the SC may be abundant nuclear proteins. Comparisons of *Xenopus laevis* spermatocyte and spermatid nuclear proteins by SDS gel electrophoresis followed by silver staining have demonstrated that there are few differences in the nonhistones of these cell types (see Fig. 4-3). Since polycomplexes have not been found in *Xenopus* spermatid nuclei (Risley, unpublished), it seems reasonable to suggest that abundant SC proteins may remain in the nucleus in an unassembled form after meiosis. Such abundant SC proteins may also be present in an unassembled form in premeiotic or somatic cell nuclei. Proteins that are unique to the SC and meiotic prophase are probably quantitatively minor and not resolveable in one dimensional SDS gels of total nuclear proteins. The presence of unassembled abundant SC proteins in postmeiotic nuclei would also explain the similarity in the protein composition of mouse pachytene spermatocyte nuclear matrices (with SCs) and spermatid nuclear matrices (which lack SCs) observed by Ierardi et al. (1983).

Cytochemical studies suggest that the SCs also may contain RNA. Using the uranyl-EDTA-lead regressive staining technique for RNA (Bernhard, 1969), Esponda and Stockert (1971, 1972), and Esponda and Giménez-Martinez (1972) have demonstrated positive staining in the lateral elements, central region, and attachment sites of SCs. RNase treatments removed the

Figure 4-3. Silver stained, SDS-acrylamide (8%) gel containing nuclear proteins from *Xenopus laevis (A)* primary spermatocytes and *(B)* early-mid stage spermatids. LI and LII indicate the positions of *Xenopus* lamins I and II.

positive staining material. In another study (Westergaard and von Wettstein, 1970), RNase was shown to remove SC material, which stained positively with ammoniacal silver. RNA associated with the SCs does not appear to be essential for the maintenance of SC structure, since RNase treatment does not disrupt SCs. The RNA may, nevertheless, be important for the proper assembly of SCs and their attachment to chromatin.

The presence of DNA in the SC is expected, since pairing and crossing over both require the interaction of homologous DNA strands. DNA has not been localized in the central components by cytochemical means, perhaps because it occurs only in very small amounts. The presence of DNA in this region is, however, suggested by the observation that ^{3}H-thymidine is incorporated near the recombination nodules (Carpenter, 1981).

Material that stains with indium trichloride (Coleman and Moses, 1964) and is removable with DNase (Coleman and Moses, 1964; Solari, 1972) is present in the lateral elements. Solari (1972) has visualized a 65 Å diameter, DNase sensitive filament in the longitudinal axes of lateral elements from rat and hamster spermatocytes that were spread and observed using whole mount and negative stain techniques. Each lateral element contained a single filament that appeared to bind the bases of chromatin loops from the attached homologue. The presence of DNA in the lateral elements is further suggested by the observation that ^{3}H-thymidine incorporated by zygotene microsporocytes from *Lilium* can be localized autoradiographically in the lateral elements (Kurata and Ito, 1978). DNA filaments in the lateral elements represent a small percent of the genome (e.g., 0.006% in lilies, 0.01% in humans, 0.3% in *Neurospora*: from von Wettstein et al., 1984), but may be very important in SC assembly and synapsis.

SYNAPTONEMAL COMPLEXES, CHROMATIN LOOPS, AND CHROMOSOME SCAFFOLDS

As mentioned above, electron microscopy of spread meiotic chromosomes has shown that the DNA in each pachytene homologue binds to the lateral elements in a looped arrangement (Comings and Okada, 1970, 1971; Solari, 1972; Kierszenbaum and Tres, 1974; Rattner et al., 1980, 1981). The chromatin loops in *Bombyx* spermatocytes form during early meiotic prophase, before the SCs are formed (Rattner et al., 1981). At that stage, the loops appear to contain 5-25 μm of DNA, and are each attached at their ends to a chromatin filament with the typical "beads-on-a-string" nucleosomal organization. Early in meiosis, the loops are separated by 1500-2000 Å; however, the loop density increases to 26 loops/μm at pachytene and 175 loops/μm at metaphase.

In pachytene spermatocytes from *Bombyx*, the chromatin loops are anchored to lateral elements of SCs in a clustered distribution (Rattner et al., 1981). This clustered arrangement of loops has been noted in other organisms, and has been correlated with the pattern of pachytene chromomeres (Comings and Okada, 1970, 1971; Comings and Okada, 1975). Since the chromatin fiber in the mitotic chromosome is also folded into numerous loops (see chap. 3), it is possible that the looped domains of mitotic and meiotic chromosomes are similar. This possibility is enhanced by the observation that the pachytene chromomere pattern is similar to the G-band pattern of mitotic chromosomes (Ferguson-Smith and Page, 1973; Okada and Comings, 1974; Luciani et al., 1975) and the loop cluster patterns of pachytene and mitotic chromosomes (Comings and Okada, 1975).

Similarities between the organization of meiotic and mitotic chromosomes can also be found in a comparison of the lateral elements of SCs to mitotic chromosome scaffolds (see chap. 3). Like the lateral elements, scaffolds are located along the longitudinal axis of chromosomes, and appear to act as attachment sites for chromatin loops (Adolph et al., 1977; Paulson and Laemmli, 1977; Marsden and Laemmli, 1979). They are fibrous structures comprised chiefly of nonhistone proteins, and are resistant to extraction with 2 M NaCl, or digestion by DNase and RNase (Adolph et al., 1977). The scaffolds also stain preferentially with silver, as do the lateral elements (Earnshaw and Laemmli, 1984).

The similarities between lateral elements and scaffolds suggest that they may be comprised of the same macromolecules. Lateral elements and scaffolds may assemble in a similar manner, by the coalescence of macromolecules located at the bases of the chromatin looped domains. The coalescence of these regions could be the cause or the result of chromosome condensation. This hypothesis is attractive since it suggests that the bases of the chromatin loops may be fundamental units of chromosome structure important for the regulation of diverse chromosome functions, such as condensation and synapsis. It should be noted, however, that in some organisms (e.g., *Lilium*: Moens, 1968; *Aedes*: Fiil and Moens, 1973; Fiil, 1978) the lateral elements appear to be removed from the chromosomes during diplotene, and persist as extrachromosomal structures in nuclei. Chromosome organization does not appear to be disrupted in such organisms, since they continue through the meiotic divisions. This suggests that the lateral elements and scaffolds are either not related structures, or that the extrachromosomal lateral elements consist of excess subunits.

SYNAPTONEMAL COMPLEXES AND THE NUCLEAR MATRIX

Several reports have suggested that SCs are integral components of a spermatocyte nuclear matrix (Comings and Okada, 1976; Gambino et al., 1981;

Ierardi et al., 1983; Raveh and Ben-Ze'ev, 1984; Risley, 1985). Nuclear matrices are residual nuclear structures resulting from the treatment of isolated nuclei with nonionic detergents, nucleases, and high salt concentrations (1-2 M KCl or NaCl) to remove chromatin and RNA (for review, see Berezney, 1979; Agutter and Richardson, 1980; Bouteille et al., 1983). The preparative procedures vary in detail, and the morphologies of the resultant structures also vary; however, matrices usually have a peripheral nuclear lamina with embedded pore complexes, and may contain residual nucleoli and variable amounts of an internal fibrogranular network. Due to the variable morphologies obtained, the nature of the nuclear matrix is controversial, and there is some concern that residual structures other than the pore complex-lamina fraction are artifacts created by salt induced aggregation or oxidation of sulfhydral groups to generate disulfide cross-links (Kaufmann et al., 1981; Kaufmann and Shaper, 1984).

Certain observations suggest that SCs may be bona fide constituents of a nuclear matrix or skeleton. First, in several species, SC components can be found as extrachromosomal structures before or after pachytene. The SCs therefore may be considered as extrachromosomal structures that play a direct role in the positioning of chromosomes in the meiocyte nucleus. Second, the total SC complement is integrated into a three-dimensional network as a result of their attachments to the nuclear membrane. The SC-membrane interactions may involve the pore-complex lamina. Third, the SCs are similar to matrix constituents in their resistance to nuclease digestion and salt extractions.

Attempts to isolate meiocyte nuclear matrices containing intact SCs have yielded conflicting observations. Walmsley and Moses (1981) were unable to isolate intact nuclear matrices from hamster pachytene spermatocytes, but did obtain intact SCs that were not a part of another structure. In contrast, Ierardi et al. (1983) have reported the isolation of nuclear matrices from mouse pachytene spermatocytes. The matrices from these cells consisted of a peripheral boundary of nuclear pores connected by a fine network of filaments, an inner fibrogranular component, and intact SCs. Nuclear matrices containing SCs have also been isolated from *Xenopus laevis* spermatocytes (Gambino et al., 1981; Risley, 1985) and from oocytes of the flour moth, *Ephestia* (Raveh and Ben-Ze'ev, 1984). The pachytene matrices described in these reports differed from those isolated from the mouse in that they consisted primarily of a peripheral pore complex-lamina and SCs, and little fibrogranular material (see Fig. 4-4). In the *Xenopus* spermatocyte matrices, the SC termini often appeared to be attached to the peripheral lamina.

The results described above may vary for any of several reasons. First, endogenous levels of proteases and RNases may vary in meiocytes from different species. *Xenopus* spermatocytes have relatively high levels of protease and RNase activities (Risley, unpublished observations). Nuclei isolated from these cells by standard methods frequently show morphological alterations within 2 hrs. These degenerative changes can be minimized by isolating the nuclei (impure) rapidly or by adding glycerol to isolation buffers and conducting the isolations at $-20°C$. Second, the inner fibrogranular component may vary in content due to different degrees of disulfide bond formation during the isolation of the matrices (Kaufmann et al., 1981; Kaufmann and Shaper, 1984). Third, since spermatocytes tend to degenerate in vitro, there

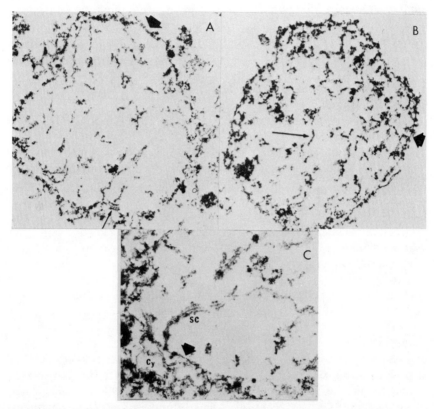

Figure 4-4. Electronmicrographs of nuclear matrices isolated from *Xenopus laevis* primary spermatocytes. Nuclei (impure) were isolated rapidly by gentle homogenization in 0.2 M sucrose, 3 mM $CaCl_2$, 10 mM Tris-HCL (pH 7.2), 1 mM PMSF, 0.5% Triton X-100. They were then pelleted, washed in 10 mM Tris-HCL (pH 7.2), 3 mM $MgCl_2$, 1 mM PMSF, and digested with 100 µg/ml DNase I for 5–10 mins. Matrices were prepared by extraction with 2 M NaCl, 3 mM $MgCl_2$, 10 mM Tris-HCL (pH 7.2). Broad arrows, nuclear lamina; long arrows and SC, synaptonemal complexes; Cy, residual cytoskeletal contaminants. Magnifications are *(A)* 11,592 x; *(B)* 10,192 x; *(C)* 12,936 x. (The nuclear matrices were isolated in collaboration with Dr. John Gambino, Department of Biology, Brooklyn College, Brooklyn, New York.)

may be species differences in the rates of spermatocyte degeneration. For example, *Xenopus* spermatocytes in culture will continue to develop through meiosis (Risley, 1983), while mammalian spermatocytes gradually degenerate.

The association of SCs with a peripheral nuclear lamina, and the in vivo attachment of cytoplasmic microfilaments to the nuclear membrane-SC attachment plaques (Solari, 1970; Esponda and Giménez-Martin, 1972), suggest that the SCs may be members of a pervasive cytoskeletal network that includes both nuclear and cytoplasmic constituents. This skeletal network could be important in providing the direction and forces required for the movements of the telomere attachment sites in the plane of the nuclear membrane during meiotic prophase. Interference by microtubule antagonists (colchicine) in synapsis and chiasma formation (Shephard et al., 1974; Bennett et al., 1979*b*) is consistent with a role of the cytoskeleton in these processes.

DNA METABOLISM DURING MEIOSIS

The molecular mechanisms of homologue synapsis, recombination, and chiasma formation are not known, but it is generally recognized that the organization and metabolism of DNA in meiotic cells are central to these processes. Strand exchange between homologues and the resolution of interlocked bivalents during zygotene both require the breakage and reunion of DNA helices. Synapsis also may depend upon the synthesis of DNA during early prophase, since chromosomes fail to synapse and SCs do not assemble in lily microsporocytes cultured in deoxyadenosine (Roth and Ito, 1967) or nalidixic acid (Takegami, 1983), both potent inhibitors of DNA synthesis. Stern and his collaborators have demonstrated several unique properties of DNA synthesis and organization in zygotene and pachytene meiocytes that may be important for pairing and recombination.

The occurrence of DNA synthesis during zygotene and pachytene was first demonstrated in lily microsporocytes (Hotta et al., 1966) and subsequently in mouse spermatocytes (Hotta et al., 1977). In lilies, DNA synthesis at zygotene is a delayed semi-conservative replication of approximately 0.3% of the genome that is not replicated during premeiotic S phase. Pachytene synthesis of DNA is predominantly of the repair type, and occurs at relatively low levels (0.1% of premeiotic S phase).

DNA synthesized at zygotene (zygDNA: Hotta et al., 1984) consists of unique or low copy number sequences that appear to be distributed throughout the genome (Hotta and Stern, 1975). Only 50% of the zygDNA present at pachytene is found in preleptotene or leptotene meiocytes, clearly demonstrating that zygDNA is delayed in its replication (Hotta and Stern, 1971). In addition to the replication delay, zygDNA sequences have the unique property of remaining unligated to neighboring DNA sequences until the end of meiotic prophase (Hotta and Stern, 1976). The zygDNA sequences are 5-10 kbp fragments flanked by gaps. As a result, these sequences can be preferentially released from pachytene nuclei or DNA by mechanical and enzymatic (S1 nuclease) treatments (Hotta and Stern, 1975, 1976). ZygDNA can also be separated from bulk DNA by centrifugation in CsCl gradients, since the zygDNA sequences in lilies are enriched in GC base pairs (50%) relative to bulk DNA (40% GC) (Hotta and Stern, 1971). It should also be noted that the zygDNA sequences are not unique to meiotic cells (Hotta and Stern, 1971).

Hotta et al. (1984) have recently investigated the delayed replication and organization of zygDNA sequences using cloned zygDNA sequences as probes. The release of zygDNA from isolated meiocyte nuclei by S1 nuclease was shown to be stage dependent. The sequences were released from pachytene and zygotene nuclei, but not from leptotene nuclei unless these nuclei were first extracted with 0.2% DOC (deoxycholate). Interestingly, DOC extraction also removed the inhibition of zygDNA synthesis in isolated leptotene nuclei. DOC extracts from leptotene or preleptotene nuclei blocked zygDNA synthesis in zygotene nuclei in vitro, but had no effect on bulk DNA synthesis in nuclei from S phase cells. Extracts of zygotene, pachytene, or somatic cell nuclei had no effect on zygDNA synthesis in isolated nuclei.

The results described above suggested that a factor extracted by DOC is bound to zygDNA in chromatin, preventing zygDNA release by S1 nuclease, and also blocking its replication (Hotta et al., 1984). At zygotene, the factor is probably removed from zygDNA, permitting the initiation of zygDNA synthesis. To identify the factor, DOC extracts were fractionated, using ammonium sulfate precipitation and DNA cellulose chromatography. A 73 kd protein, termed L protein, was isolated and shown to have the properties of a zygDNA binding protein capable of specific inhibition of zygDNA synthesis (Hotta et al., 1984). L protein binds to a 90 bp region in zygDNA, and preferentially binds relaxed or negatively supercoiled DNA.

The ability of deoxyadenosine to inhibit zygDNA synthesis and block synapsis and SC formation suggests that zygDNA synthesis is important to meiosis (Roth and Ito, 1967). This is reinforced by the observation that zygDNA synthesis is associated with the SCs (Kurata and Ito, 1978). Inhibition of zygDNA synthesis by hydroxyurea, however, does not prevent SC formation in lilies (Takegami and Ito, 1982). Both Hotta et al. (1984) and Takegami and Ito (1982) have speculated that the initiation of zygDNA synthesis at the appropriate time is the most important aspect of zygDNA metabolism; zygDNA synthesis must remain suppressed until all of the necessary conditions for synapsis have been established. Once initiated, inhibition of zygDNA synthesis may not fully block synapsis. Consistent with this interpretation, Hotta et al. (1984) have shown that L protein activity is present only in preleptotene meiocytes that are committed to meiosis. Premeiotic S cells, which revert to the mitotic cycle when explanted and cultured, lack L protein activity. In these cells, zygDNA is replicated before synapsis can occur.

Synthesis of DNA during pachytene occurs in a portion of the genome (P-DNA), which is distinct from the zygDNA. P-DNA synthesis is a repair synthesis associated with sequences that are nicked by a meiosis specific endonuclease during late zygotene or early pachytene (Howell and Stern, 1971). The average distance between the nicks in lily meiocytes is 160 kbp, and the nicks occur in moderately repetitive sequences that are repeated about 1000 times per genome in lilies (Hotta and Stern, 1974) and 400 times in mouse (Stern and Hotta, 1977). There are approximately 600 families of the P-DNA repeats in lily, with each family containing sequences 1500-2000 bp in length repeated about 1000 times (Bouchard and Stern, 1980). Unlike other moderately repetitive sequences, each P-DNA sequence family exhibits very low divergence. Moreover, there appears to be considerable (94%) sequence conservation in the P-DNA of higher plants, although the repeat frequencies vary (Friedman et al., 1982).

P-DNA synthesis occurs within a 100-200 bp subset of the P-DNA repeat unit, as assayed by ^{3}H-thymidine incorporation. Hotta and Stern (1984) have recently isolated the P-DNA repeats from lily meiocyte nuclei by taking advantage of their sensitivity to digestion by DNase II (Hotta and Stern, 1981). The repeat lengths range from 800-3000 bp, and contain three discrete regions. At each end there are 150-300 bp sequences termed (PsnDNA) (Psn = pachytene small nuclear) that are the sites of nicking and repair synthesis during pachytene. A lower copy number sequence with no homology to the PsnDNAs is located in the middle of the P-DNA repeat. PsnDNA nicks appear to be introduced in about half of the genome with a spacing of 30-350 kbp. The nicking is therefore not random.

The nick and repair process coincides with crossing over, and with large increases in the activities of the meiotic endonuclease, as well as the meiosis specific U (unwinding) and R (reannealing) proteins hypothesized to be important for crossing over (see Stern, 1981). Initiation of P-DNA nick and repair synthesis is likely to be dependent upon the increase in activity of the endonuclease; however, it is also dependent upon synapsis, since endonuclease levels are normal in the achiasmate lily hybrid, Black Beauty, but nicking and P-DNA synthesis do not occur (Hotta et al., 1979). This suggests that there may be a difference in the organization of P-DNA in the chromatin of chiasmate and achiasmate cells, and that the accessibility of the P-DNA plays a role in the control of the nicking process (Hotta et al., 1979). The observation that P-DNA in lily meiocyte nuclei is preferentially sensitive to micrococcal nuclease digestion is consistent with this suggestion (Stern and Hotta, 1977). Hotta and Stern (1981) have isolated from lily meiocytes a 125 nucleotide small nuclear RNA (PsnRNA) that appears to be important in regulating the accessibility of the P-DNA sequences to meiotic endonuclease or DNase II.

PsnRNA was isolated as a component of an RNase resistant, Mg^{++}-soluble chromatin fraction released from pachytene nuclei by DNase II. The chromatin fragments contained 85-95% of the P-DNA, but only 5-10% of the total DNA. The PsnRNA was about 10% of the total nuclear RNA, and one third of the RNA synthesized at pachytene. Little, if any, PsnRNA was found in prezygotene or postpachytene nuclei. Based on α-amantin sensitivity, synthesis of PsnRNA appears to be performed by RNA polymerase III.

A direct role for PsnRNA in the regulation of P-DNA chromatin organization was suggested by the observation that the nuclease insensitive P-DNA sequences in the achiasmate pachytene nuclei of the Black Beauty hybrid were rendered accessible to meiotic endonuclease or DNase II when PsnRNA was added to these nuclei in vitro (Hotta and Stern, 1981). Treatment of cross-over proficient lily pachytene meiocytes with α-amanitin caused the P-DNA sequences to lose their preferential sensitivity to nuclease digestion. Interestingly, PsnRNA added to preleptotene or leptotene nuclei had little effect on P-DNA accessibility.

The role, if any, of P-DNA nick and repair activity in meiosis is not clear. Temporal aspects of the process, its dependence upon effective synapsis, and its absence from achiasmate hybrids, suggest that P-DNA nicking and repair synthesis may be important in crossing over and chiasma formation. The number of sites of nick and repair, however, are enormous (650,000: Bouchard and Stern, 1980) in comparison to the number of crossovers that occur (1-3 per bivalent). Most sites may represent potential crossover domains, of which only a fraction successfully participate in chiasma formation.

ACKNOWLEDGMENTS

This work was supported, in part, by grant 1 R23 HD16154 from the National Institute of Child Health and Human Development, and by grant 1 RO1 ESHD03381 from the National Institute of Environmental Health Sciences.

REFERENCES

Adolph, K. W.; Cheng, S. M.; and Laemmli, U. K. (1977). Role of nonhistone proteins in metaphase chromosome structure. *Cell* **12,** 805-816.
Agutter, P. S.; and Richardson, J. C. W. (1980). Nuclear non-chromatin proteinaceous structures: their role in the organization and function of the interphase nucleus. *J. Cell Sci.* **44,** 395-435.

Baker, B. S.; Carpenter, A. T. C.; Esposito, M. S.; Esposito, R. E.; and Sandler, L. (1976). The genetic control of meiosis. *Annu. Rev. Genet.* **10,** 53-134.

Bennett, M. D.; Smith, J. B.; Simpson, S.; and Wells, B. (1979a). Intranuclear fibrillar material in cereal pollen mother cells. *Chromosoma* **71,** 289-332.

Bennett, M. D.; Toledo, L. A.; and Stern, H. (1979b). The effect of colchicine on meiosis in Lilium speciosum cv. "Rosemede." *Chromosoma* **72,** 175-189.

Berezney, R. (1979). Dynamic properties of the nuclear matrix. In *The Cell Nucleus* (H. Busch, ed.), vol. 7, pp. 413-456, Academic Press, New York.

Bernhard, W. (1969). A new staining procedure for electron microscopical cytology. *J. Ultrastruct. Res.* **27,** 250-265.

Bogdanov, Yu. F. (1977). Formation of cytoplasmic synaptone-mal-like polycomplexes at leptotene and normal synaptonemal complexes at zygotene in *Ascaris suum* male meiosis. *Chromosoma* **61,** 1-21.

Bojko, M. (1983). Human meiosis VIII. Chromosome pairing and formation of the synaptonemal complex in oocytes. *Carlsberg Res. Commun.* **48,** 457-483.

Bouchard, R. A.; and Stern, H. (1980). DNA synthesized at pachytene in *Lilium*: A nondivergent subclass of moderately repetitive sequences. *Chromosoma* **81,** 349-363.

Bouteille, M.; Bouvier, D.; and Seve, A. P.; (1983). Heterogeneity and territorial organization of the nuclear matrix and related structures. *Int. Rev. Cytol.* **83,** 135-182.

Callan, H. G. (1972). Replication of DNA in the chromosomes of eukaryotes. *Proc. R. Soc. Lond.* **B181,** 19-41.

Callan, H. G.; and Taylor, J. H. (1968). A radioautographic study of the time course of male meiosis in the newt *Triturus vulgaris*. *J. Cell Sci.* **3,** 615-626.

Carmi, P.; Koltin, Y.; and Stamberg, J. (1978). Meiosis in *Schizophyllum commune*: Premeiotic DNA replication and meiotic synchrony induced with hydroxyurea. *Genet. Res.* **31,** 215-226.

Carpenter, A. T. C. (1975). Electron microscopy of meiosis in *Drosophila melanogaster* females. II. The recombination nodule—a recombination associated structure at pachytene? *Proc. Nat. Acad. Sci. (U.S.A.)* **72,** 3186-3189.

Carpenter, A. T. C. (1979). Recombination nodules and synaptonemal complex in recombination defective females of *Drosophila melanogaster*. *Chromosoma* **75,** 259-292.

Carpenter, A. T. C. (1981). EM autoradiographic evidence that DNA synthesis occurs at recombination nodules during meiosis in *Drosophila melanogaster* females. *Chromosoma* **83,** 59-80.

Coleman, J. R.; and Moses, M. J. (1964). DNA and the fine structure of synaptic chromosomes in the domestic rooster (*Gallus domesticus*). *J. Cell Biol.* **23,** 63-78.

Comings, D. E. (1980). Arrangement of chromatin in the nucleus. *Hum. Genet.* **53,** 131-143.

Comings, D. E.; and Okada, T. A. (1970). Whole mount electron microscopy of meiotic chromosomes and the synaptonemal complex. *Chromosoma* **30,** 269-286.

Comings, D. E.; and Okada, T. A. (1971). Whole mount electron microscopy of human meiotic chromosomes. *Exp. Cell Res.* **65,** 99-103.

Comings, D. E.; and Okada, T. A. (1975). Mechanisms of chromosome banding. VI. Whole mount electron microscopy of banded metaphase chromosomes and a comparison with pachytene chromosomes. *Exp. Cell Res.* **93,** 267-274.

Comings, D. E.; and Okada, T. A. (1976). Nuclear proteins. III. The fibrillar nature of the nuclear matrix. *Exp. Cell Res.* **103,** 341-360.

Counce, S. J.; and Meyer, G. F. (1973). Differentiation of the synaptonemal complex and the kinetochore in *Locusta* spermatocytes studied by whole mount electron microscopy. *Chromosoma* **44,** 231-253.

De Martino, C.; Capanna, E.; Nicotra, M. R.; and Natali, P. G. (1980). Immunochemical localization of contractile proteins in mammalian meiotic chromosomes. *Cell Tissue Res.* **213**, 159-178.

Dresser, M. E.; and Moses, M. J. (1979). Silver staining of synaptonemal complexes in surface spreads for light microscopy and electron microscopy. *Exp. Cell Res.* **121**, 416-419.

Dresser, M. E.; and Moses, M. J. (1980). Synaptonemal karyotyping in spermatocytes of the Chinese hampster (*Cricetulus griseus*). IV. Light and electron microscopy of synapsis and nucleolar development by silver staining. *Chromosoma* **76**, 1-22.

Dresser, M. E.; and Moses, M. J. (1983). Cytological characterization of an anti-synaptonemal complex monoclonal antibody. *J. Cell Biol.* **97**, 185a.

Earnshaw, W. C.; and Laemmli, U. K. (1984). Silver staining the chromosome scaffold. *Chromosoma* **89**, 186-192.

Esponda, P.; and Giménez-Martin, G. (1972). The attachment of the synaptonemal complex to the nuclear envelope. An ultrastructural and cytochemical analysis. *Chromosoma* **38**, 405-417.

Esponda, P.; and Krimer, D. B. (1979). Development of the synaptonemal complex and polycomplex formation in three species of grasshoppers. *Chromosoma* **73**, 237-245.

Esponda, P.; and Stockert, J. C. (1971). Localization of RNA in the synaptonemal complex. *J. Ultrastruct. Res.* **35**, 411-417.

Esponda, P.; and Stockert, J. C. (1972). Evolution of the synaptonemal complex in *Helix aspersa* spermatocytes. *Chromosoma* **36**, 150-157.

Fawcett, D. W. (1956). The fine structure of chromosomes in the meiotic prophase of vertebrate spermatocytes. *J. Biophys. Biochem. Cytol.* **2**, 403-406.

Ferguson-Smith, M. A.; and Page, B. M. (1973). Pachytene analysis in a human reciprocal (10:11) translocation. *J. Med., Genet.* **10**, 282-287.

Fiil, A. (1978). Meiotic chromosome pairing and synaptonemal complex transformation in *Culex pipiens* oocytes. *Chromosoma* **69**, 381-395.

Fiil, A.; and Moens, P. B. (1973). The development, structure and function of modified synaptonemal complexes in mosquito oocytes. *Chromosoma* **36**, 119-130.

Fiil, A.; Goldstein, P.; and Moens, P. B. (1977). Precocious formation of synaptonemal-like polycomplexes and their subsequent fate in female *Ascaris lumbricoides* var. *suum*. *Chromosoma* **65**, 21-35.

Fletcher, J. M. (1979). Light microscope analysis of meiotic prophase chromosomes by silver staining. *Chromosoma* **72**, 241-248.

Friedman, B. E.; Bouchard, R. A.; and Stern, H. (1982). DNA sequences repaired at pachytene exhibit strong homology among distantly related higher plants. *Chromosoma* **87**, 409-424.

Gambino, J.; Eckhardt, R. A.; and Risley, M. S. (1981). Nuclear matrices containing synaptonemal complexes from *Xenopus laevis*. *J. Cell Biol.* **99**, 63a.

Gillies, C. B. (1972). Reconstruction of the *Neurospora crassa* pachytene karyotype from serial sections of synaptonemal complexes. *Chromosoma* **36**, 119-130.

Gillies, C. B. (1973). Ultrastructural analysis of maize pachytene karyotypes by three dimensional reconstruction of the synaptonemal complexes. *Chromosoma* **43**, 145-176.

Gillies, C. B. (1974). The nature and extent of synaptonemal complex formation in haploid barley. *Chromosoma* **48**, 441-453.

Gillies, C. B. (1979). The relationship between synaptonemal complexes recombination nodules and crossing over in *Neurospora crassa* bivalents and translocation quadrivalents. *Genetics* **91**, 1-17.

Golubovskaya, I. N. (1979). Genetic control of meiosis. *Int. Rev. Cytol.* **58**, 247-290.

Grell, R. F. (1978). Time of recombination in the Drosophila melanogaster oocyte: evidence from a temperature-sensitive recombination-deficient mutant. *Proc. Natl. Acad. Sci. (U.S.A.)* **75**, 3351-3354.

Grell, R. F.; and Generoso, E. E. (1980). Time of recombination in the *Drosophila melanogaster* oocyte. II. Electron microscopic and genetic studies of a temperature sensitive recombination mutant. *Chromosoma* **81,** 339-349.

Grell, R. F.; and Generoso, E. E. (1982). A temporal study at the ultrastructural level of the developing pro-oocyte of *Drosophila melanogaster. Chromosoma* **87,** 49-75.

Grell, R. F.; Oakberg, E. F.; and Generoso, E. E. (1980). Synaptonemal complexes at premeiotic interphase in the mouse spermatocyte. *Proc. Nat. Acad. Sci. (U.S.A.)* **77,** 6720-6723.

Henderson, S. A. (1970). The time and place of meiotic crossing over. *Annu. Rev. Genet.* **4,** 295-324.

Holm, P. B. (1977a). The premeiotic DNA replication of euchromatin in *Lilium longiflorum* (Thunb.). *Carlsberg Res. Commun.* **42,** 249-281.

Holm, P. B. (1977b). Three dimensional reconstruction of chromosome pairing during the zygotene stage of meiosis in *Lilium longiflorum* (Thunb.). *Carlsberg Res. Commun.* **42,** 103-151.

Holm, P. B.; and Rasmussen, S. W. (1980). Chromosome pairing, recombination nodules and chiasma formation in diploid *Bombyx* males. *Carlsberg Res. Commun.* **45,** 483-548.

Holm, P. B.; Rasmussen, S. W.; Zickler, D.; Lu, B. C.; and Sage, J. (1981). Chromosome pairing, recombination nodules and chiasma formation in the basidiomycete *Coprinus cinereus. Carlsberg Res. Commun.* **46,** 305-346.

Hotta, Y.; and Stern, H. (1971). Analysis of DNA synthesis during meiotic prophase in *Lilium. J. Mol. Biol.* **55,** 337-355.

Hotta, Y.; and Stern, M. (1974). DNA scission and repair during pachytene in *Lilium. Chromosoma* **46,** 279-296.

Hotta, Y.; and Stern, H. (1975). Zygotene and pachytene-labeled sequences in the meiotic organization of chromosomes. In *The Eukaryotic Chromosome* (N. S. Peacock and R. D. Brock, eds.), pp. 283-300, Australian National University Press, Canberra.

Hotta, Y.; and Stern, H. (1976). Persistent discontinuities in late replicating DNA during meiosis in *Lilium. Chromosoma* **55,** 171-182.

Hotta, Y.; and Stern, H. (1981). Small nuclear RNA molecules that regulate nuclease accessibility in specific chromatin regions of meiotic cells. *Cell* **27,** 309-319.

Hotta, Y.; and Stern, H. (1984). The organization of DNA segments undergoing repair synthesis during pachytene. *Chromosoma* **89,** 127-137.

Hotta, Y.; Ito, M.; and Stern, H. (1966). Synthesis of DNA during meiosis. *Proc. Nat. Acad. Sci. (U.S.A.)* **56,** 1184-1191.

Hotta, Y.; Chandley, A.; and Stern, H. (1977) Biochemical analysis of meiosis in the male mouse. II. DNA metabolism at pachytene. *Chromosoma* **62,** 255-268.

Hotta, Y.; Bennett, M. D.; Toledo, L. A.; and Stern, H. (1979). Regulation of R-protein and endonuclease activities in meiocytes by homologous chromosome pairing. *Chromosoma* **72,** 191-201.

Hotta, Y.; Tabata, S.; and Stern, H. (1984). Replication and nicking of zygotene DNA sequences. *Chromosoma* **90,** 243-253.

Howell, S. H.; and Stern, H. (1971). The appearance of DNA breakage and repair activities in the synchronous meiotic cycle of *Lilium. J. Mol. Biol.* **55,** 357-378.

Ierardi, L. A.; Moss, S. B.; and Bellvé, A. R. (1983). Synaptonemal complexes are integral components of the isolated mouse spermatocyte nuclear matrix. *J. Cell Biol.* **96,** 1717-1726.

Ito, M.; and Hotta, Y. (1973). Radioautography of incorporated ^{3}H-thymidine and its metabolism during meiotic prophase in microsporocytes of *Lilium. Chromosoma* **43,** 391-398.

Kaufman, S. H.; and Shaper, J. H. (1984). A subset of non-histone nuclear proteins reversibly stabilized by the sulfhydral cross-linking reagent tetrathionate. *Exp. Cell Res.* **155,** 477-495.

Kaufman, S. H.; Coffey, D. S.; and Shaper, J. H. (1981). Considerations in the isolation of rat liver nuclear matrix, nuclear envelope, and pore complex lamina. *Exp. Cell Res.* **132,** 105-123.

Kehlhoffner, J. L.; and Dietrich, J. (1983). Synaptonemal complex and a new type of nuclear polycomplex in three higher plants: *Paeonia teninfolia, Paeonia delavayi,* and *Nadiscantia paludosa. Chromosoma* **88,** 164-170.

Kierszenbaum, A. L.; and Tres, L. L. (1974). Transcription sites in spread meiotic prophase chromosomes from mouse spermatocytes. *J. Cell Biol.* **63,** 923-935.

King, R. (1970). The meiotic behavior of the Drosophila oocyte. *Int. Rev. Cytol.* **28,** 125-168.

King, R. C.; and Akai, H. (1971). Spermatogenesis in Bombyx mori. II. The ultrastructure of synapsed bivalents. *J. Morphol.* **134,** 181-194.

Kundu, S. C.; and Bogdanov, Yu. F. (1979). Ultrastructural studies of late meiotic prophase nuclei of spermatocytes in *Ascaris suum. Chromosoma* **70,** 375-384.

Kurata, N.; Ito, M. (1978). Electron microscope autoradiography of ^{3}H-thymidine incorporation during the zygotene stage in microsporocytes of lily. *Cell Struct. Funct.* **3,** 349-356.

Li, S.; Meistrich, M. L.; Brock, W. A.; Hsu, T. C.; and Kuo, M. T. (1983). Isolation and preliminary characterization of the synaptonemal complex from rat pachytene spermatocytes. *Exp. Cell Res.* **144,** 63-72.

Lindsley, D. L.; and Sandler, L. (1977). The genetic analysis of meiosis in female *Drosophila melanogaster. Philos. Trans. R. Soc., Lond.* **B277,** 295-312.

Luciani, J. M.; Morazzini, J. R.; and Stahl, A. (1975). Identification of pachytene bivalents in human male meiosis using G-banding technique. *Chromosoma* **52,** 275-282.

McQuade, H. A.; and Pickles, D. G. (1980). Observations on synaptonemal complexes in premeiotic interphase of wheat *Triticum aestivum* Chinese Spring. *Am. J. Bot.* **67,** 1361-1373.

Marsden, M. P. F.; and Laemmli, U. K. (1979). Metaphase chromosome structure: evidence for a radial loop model. *Cell* **17,** 849-858.

Martini, G.; and Bozzini, A. (1966). Radiation induced asynaptic mutations in Durum wheat (*Triticum Durum* Desf.). *Chromosoma* **20,** 251-266.

Moens, P. B. (1968). The structure and function of the synaptonemal complex in *Lilium longiflorum* sporocytes. *Chromosoma* **23,** 418-451.

Moens, P. B. (1969a). The fine structure of meiotic chromosome polarization and pairing in *Locusta migratoria* spermatocytes. *Chromosoma* **28,** 1-25.

Moens, P. B. (1969b). Multiple core complexes in grasshopper spermatocytes and spermatids. *J. Cell Biol.* **40,** 542-551.

Moens, P. B. (1970). The fine structure of meiotic chromosome pairing in natural and artificial *Lilium* polyploids. *J. Cell Sci.* **7,** 55-64.

Moens, P. B.; and Church, K. (1979). The distribution of synaptonemal complex material in metaphase I bivalents of *Locusta* and *Chloealtis* (Orthoptera: Acrididae). *Chromosoma* **73,** 247-254.

Moens, P. B.; and Rappaport, E. (1971). Synaptic structure in the nucleus of sporulating yeast. *J. Cell Sci.* **9,** 665-677.

Monesi, V. (1962). Autoradiographic study of DNA synthesis and the cell cycle in spermatogonia and spermatocytes of the mouse testis using tritiated thymidine. *J. Cell Biol.* **14,** 1-18.

Moses, M. J. (1956). Chromosomal structures in crayfish spermatocytes. *J. Biophys. Biochem. Cytol.* **2,** 215-218.

Moses. M. J. (1968). Synaptinemal complex. *Annu. Rev. Genet.* **2,** 363-412.

Moses, M. J. (1977a). Microspreading and the synatonemal complex in cytogenetic studies. In *Chromosomes Today* (A. de la Chapelle and M. Sorsa, eds.), vol. 6, pp. 71-82, Amsterdam/New York/Oxford: Elsevier/North Holland Biomedical Press.

Moses, M. J. (1977b). Synaptonemal karyotyping in spermatocytes of the Chinese hamster (*Cricetulus griseus*). II. Morphology of the XY pair in spread preparations. *Chromosoma* **60,** 127-137.

Moses, M. J. (1985). The synaptonemal complex in meiosis. In *Proceedings of Symposium on Aneuploidy: Etiology and Mechanisms.* In Press.

Moses, M. J.; and Coleman, J. R. (1964). Structural patterns and the functional organization of chromosomes. In *The Role of Chromosomes in Development* (M. Locke, ed.), pp. 11-49, Academic Press, New York.

Moses, M. J.; and Counce, S. J. (1974). Synaptonemal complex karyotyping in spreads of mammalian spermatocytes. In *Mechanisms of Recombination* (R. F. Grell, ed.), pp. 385-390, Plenum, New York.

Moses, M. J.; and Poorman, P. A. (1981). Synaptonemal complex analysis of mouse chromosomal rearrangements. II. Synaptic adjustment in a tandem duplication. *Chromosoma* **81,** 519-535.

Moses, M. J.; Counce, S. J.; and Paulson, D. F. (1975). Synaptonemal complex complement of man in spreads of spermatocytes, with details of the sex chromosome pair. *Science* **187,** 363-365.

Okada, T. A.; and Comings, D. E. (1974). Mechanisms of chromosome banding. III. Similarity between G-bands of mitotic chromosomes and chromomeres of meiotic chromosomes. *Chromosoma* **48,** 65-71.

Pathak, S.; and Hsu, T. C. (1979). Silver-stained structures in mammalian meiotic prophase. *Chromosoma* **70,** 195-203.

Paulson, J. R.; and Laemmli, U. K. (1977). The structure of histone-depleted metaphase chromosomes. *Cell* **12,** 817-828.

Rasmussen, S. W. (1973). Ultrastructural studies of spermatogenesis *Drosophila melanogaster. Z. Zellforsch.* **140,** 125-144.

Rasmussen, S. W. (1975a). Ultrastructural studies of meiosis in males and females of the c(3)G[17] mutant of *Drosophila melanogaster. C. R. Trav. Lab. Carlsberg* **40,** 163-173.

Rasmussen, S. W. (1975b). Synaptonemal polycomplexes in *Drosophila melanogaster. Chromosoma* **49,** 321-331.

Rasmussen, S. W. (1976). The meiotic prophase in *Bombyx mori* females analysed by three dimensional reconstructions of synaptonemal complexes. *Chromosoma* **54,** 245-293.

Rasmussen, S. W. (1977a). The transformation of the synaptonemal complex into the "elimination chromatin" in *Bombyx mori* oocytes. *Chromosoma* **60,** 205-221.

Rasmussen, S. W. (1977b). Chromosome pairing in triploid females of *Bombyx mori* analysed by three dimensional reconstructions of synaptonemal complexes. *Carlsberg Res. Commun.* **42,** 163-197.

Rasmussen, S. W.; and Holm, P. B. (1978a). Human meiosis. II. chromosome pairing and recombination nodules in human spermatocytes. *Carlsberg Res. Commun.* **43,** 275-327.

Rasmussen, S. W.; and Holm, P. B. (1978b). Human meiosis. IV. The elimination of synaptonemal complex fragments from metaphase I bivalents of human spermatocytes. *Carlsberg Res. Commun.* **43,** 423-438.

Rasmussen, S. W.; and Holm, P. B. (1980). Mechanics of meiosis. *Hereditas* **93,** 187-16.

Rattner, J. B.; Goldsmith, M.; and Hamkalo, B. A. (1980). Chromatin organization during meiotic prophase of *Bombyx mori. Chromosoma* **79,** 215-224.

Rattner, J. B.; Goldsmith, M.; and Hamkalo, B. A. (1981). Chromosome organization during male meiosis in *Bombyx mori. Chromosoma* **82,** 341-351.

Raveh, D.; and Ben-Zêev, A. (1984). The synaptonemal complex as part of the nuclear matrix of the flour moth, *Ephestia kuehniella. Exp. Cell Res.* **153,** 99-108.

Risley, M. S. (1983). Spermatogenic cell differentiation *in vitro. Gamete Res.* **4,** 331-346.

Risley, M. S. (1985). Changes in the nuclear lamina during spermatogenesis in *Xenopus laevis. J. Cell Biochem.*, supp. 9A, p. 10.

Roth, T. F. (1966). Changes in the synaptonemal complex during meiotic prophase in mosquito oocytes. *Protoplasma* **61,** 346-386.

Roth, T. F.; and Ito, M. (1967). DNA dependent formation of the synaptonemal complex at meiotic prophase. *J. Cell Biol.* **35,** 247-255.

Shephard, J.; Boothroyd, E. R.; and Stern, H. (1974). The effect of colchicine on synapsis and chiasma formation in microsporocytes of Lilium. *Chromosoma* **44**, 423-437.

Sheridan, W. F.; and Barnett, R. J. (1969). Cytochemical studies on chromosome ultrastructure. *J. Ultrastruct. Res.* **27**, 216-229.

Smith, P. A.; and King, R. C. (1968). Genetic control of synaptonemal complexes in *Drosophila melanogaster. Genetics* **60**, 335-351.

Solari, A. J. (1970). The spatial relationship of the X and Y chromosomes during meiotic prophase in mouse spermatocytes. *Chromosoma* **29**, 217-236.

Solari, A. J. (1972). Ultrastructure and composition of the synaptonemal complex in spread and negatively stained spermatocytes of the golden hamster and albino rat. *Chromosoma* **39**, 237-263.

Solari, A. J. (1974). The relationship between chromosomes and axes in the chiasmatic XY pair of the Armenian hamster (*Cricetulus migratorius*). *Chromosoma* **48**, 89-106.

Solari, A. J. (1977). Ultrastructure of the synaptic autosomes and the ZW bivalent in chickes oocytes. *Chromosoma* **64**, 155-165.

Solari, A. J. (1980). Synaptonemal complexes and associated structures in microspread human spermatocytes. *Chromosoma* **81**, 315-337.

Solari, A. J.; and Moses, M. J. (1973). The structure of the central region in the synaptonemal complexes of hamster and cricket spermatocytes. *J. Cell Biol.* **56**, 145-152.

Solari, A. J.; and Vilar, O. (1978). Multiple complexes in human spermatocytes. *Chromosoma* **66**, 331-340.

Sotelo, J. R.; Gracia, R. B.; and Wettstein, R. (1973). Serial sectioning study of some meiotic stages in *Scaptericus borelli* (Grylloidea). *Chromosoma* **42**, 307-334.

Stein, H. (1981). Chromosome organization and DNA metabolism in meiotic cells. In *Chromosomes Today* (M. D. Bennett, M. Bobrow, and G. Hewitt, eds.), vol. 7, pp. 94-104, George Allen and Unwin, London.

Stern, H.; and Hotta, Y. (1977). Biochemistry of meiosis. *Philos. Trans. R. Soc. Lond.* **B277**, 277-294.

Stern, H.; and Hotta, Y. (1978). Regulatory mechanisms in meiotic crossing-over. *Annu. Rev. Plant Physiol.* **29**, 415-436.

Stern, H.; Westergaard, M.; and Von Wettstein, D. (1975). Presynaptic events in meiocytes of *Lilium longiflorum* and their relation to crossing-over: A preselection hypothesis. *Proc. Natl. Acad. Sci. (U.S.A.)* **72**, 961-965.

Stringham, G. R. (1970). A cytogenetic analysis of three asynaptic mutants in *Brassica campestris* L. *Can. J. Genet. Cytol.* **12**, 743-749.

Stubblefield, E. (1973). The structure of mammalian chromosomes. *Int. Rev. Cytol.* **35**, 1-60.

Swaminathan, M. S.; and Murty, B. R. (1959). Aspects of asynapsis in plants. I. Random and non-random chromosome associations. *Genetics* **44**, 1271-1280.

Takegami, M. H. (1983). Induction of univalent chromosomes in explanted lily microsporocytes by inhibitors of DNA synthesis. *Cell Struct. Funct.* **8**, 43-55.

Takegami, M. H.; and Ito, M. (1982). Effect of hydroxyurea on mitotic and meiotic divisions in explanted lily microsporcytes. *Cell Struct. Funct.* **7**, 29-38.

Tres, L. L. (1977). Extensive pairing of the XY bivalent in mouse spermatocytes as visualized by whole-mount electron microscopy. *J. Cell Sci.* **25**, 1-15.

Wahrman, J. (1981). Synaptonemal complexes—origin and fate. In *Chromosomes Today* (M. D. Bennett, M. Bobrow, and G. Hewitt, eds.), vol. 7, pp. 105-113, George Allen and Unwin, London.

Walmsley, M.; and Moses, M. J. (1981). Isolation of synaptonemal complexes from hamster spermatocytes. *Exp. Cell Res.* **133**, 405-411.

Weith, A.; and Traut, W. (1980). Synaptonemal complexes with associated chromatin in a moth, *Ephestia kuehniella* Z. Fine structure of the W chromosomal heterochromatin. *Chromosoma* **78**, 275-291.

Westergaard, M.; and von Wettstein, D. (1970). Studies on the mechanism of crossing over. IV. The molecular organization of the synaptonemal complex in *Neotiella* (Cooke) Saccardo (Ascomycetes). *C. R. Trav. Lab. Carlsberg* **37**, 239-268.

Westergaard, M.; and von Wettstein, D. (1972). The synaptonemal complex. *Annu. Rev. Genet.* **6**, 71-110.

Wettstein, D. von; (1977). The assembly of the synaptonemal complex. *Philos. Trans. R. Soc. Lond.* **B277**, 235-243.

Wettstein, D. von; Rasmussen, S. W.; and Holm, P. B. (1984). The synaptonemal complex in genetic segregation. *Annu. Rev. Genet.* **18**, 331-413.

Wettstein, R.; and Sotelo, J. R. (1967). Electron microscope serial reconstruction of the spermatocyte I nuclei at pachytene. *J. Microscopie* **6**, 557-576.

Woolam, D. H. M.; Millen, J. W.; and Ford, E. H. R. (1967). Points of attachment of pachytene chromosomes to the nuclear membrane in mouse spermatocytes. *Nature* (Lond.) **213**, 298-299.

Zickler, D. (1973). Fine structure of chromosome pairing in ten Ascomycetes. Meiotic and premeiotic (mitotic) synaptonemal complexes. *Chromosoma* **40**, 401-416.

Zickler, D. (1977). Development of the synaptonemal complex and the "recombination nodules" during meiotic prophase in the seven bivalents of the fungus *Sordaria macrospora* Auersw. *Chromosoma* **61**, 289-316.

Zickler, D.; and Sage, J. (1981). Synaptonemal complexes with modified lateral elements in *Sordaria humana*: development of and relationship to the "recombination nodules." *Chromosoma* **84**, 305-318.

5

The Lampbrush Chromosomes of Animal Oocytes

Herbert C. Macgregor
University of Leicester
Leicester, England

Lampbrush chromosomes (LBCs) were discovered a little over 100 years ago (Flemming, 1882), and the centenary was appropriately marked by H. G. Callan's Croonian Lecture to The Royal Society (Callan, 1982). Since their discovery there have been several waves of progress in our understanding of them. Each wave has followed quickly upon the development of a new technological approach and the emergence of a new line of questioning about chromosome organization. The first series of observations on LBCs in the late nineteenth century involved relatively simple approaches through micromanipulation and light microscopy. The investigators wished to see what was inside the large swollen nuclei (germinal vesicles or GV) of animal oocytes, and they were concerned to relate what they found to the role of germ cell nuclei in heredity and development.

Much later, in the 1940s, the second wave of progress was initiated by Duryee (1937, 1941) and later gathered force through the efforts and ingenuity of Callan and J. G. Gall. The significant technological advances were in the development of phase contrast and electron microscopy, and the questions being asked were based on genetics and a well-founded understanding of chromosome behavior in meiosis. It was known that oocytes were in prophase of the first meiotic division, so their chromosomes might be expected to conform with the events that normally take place at this stage in spermatocytes. It was known that DNA was a carrier of genetic information, so it was appropriate to ask how this material was distributed along an LBC.

It was known that LBCs developed from chromatin that was packaged into a specific number of condensed metaphase chromosomes during the last oogonial mitosis, and that at the end of oogenesis the chromatin was once again condensed into the same specific array of metaphase chromosomes. So LBCs clearly represented a transitory, very strange form that was an obvious and challenging target for investigation.

The third wave of progress came with the introduction of radioisotope techniques and autoradiography into cytology, and it was boosted by a firm confidence in DNA as a source of encoded heritable and expressible information. Attention turned to the functional and structural relations between loops and chromomeres, the synthesis of RNA on loops, and the possible role of LBCs in providing materials to satisfy the requirements for development and differentiation of the embryo. It was during this phase of lampbrushology that two important ideas—the Spinning Out and Retraction Theory (Callan and Lloyd, 1960) and the Master:Slave Theory (Callan, 1967)—were generated, and although neither of them stood up to experimental test, both represented essential steps in the sequence of discovery that has led to our present rather good level of understanding.

The latest developments in lampbrush research are mainly products of observations and experiments that involved sophisticated electron microscopy and nuclei acid hybridization employing specific DNA sequences that were prepared with recombinant DNA technology. The "Miller spreading" technique for visualizing transcriptionally active chromatin was actually developed on material from amphibian germinal vesicles. In situ hybridization of labeled DNA to complementary cellular RNA—a method now widely used in the study of gene expression—was first successfully used on LBCs (Pukkila, 1975), and remains one of the most productive approaches in this field of study. Interest is now firmly focused at the molecular level, and questions relate to the arrangement and expression of DNA sequences on LBCs and the usefulness of the chromosomes as an experimental system for studying certain fundamental aspects of transcription and the regulation of gene activity.

There are lessons to be learned from the whole history of lampbrushology. We see how progress in biological science is closely linked with technological advance. The big breakthroughs on lampbrushes came with things like phase contrast microscopy and biotechnology. We see how one small system can provide a gateway to a clearer understanding of a whole range of important biological events. Enzyme digestion experiments on lampbrushes (Callan and Macgregor, 1958; Macgregor and Callan, 1962; Gall, 1963) taught us that chromosomes are enormously long, continuous, and complex molecules of DNA. LBCs are currently a valuable source of information on transcription, not to mention the many other areas where they provide unique opportunities for answering specific questions about chromosome organization, function, and evolution. Above all, LBCs are exceedingly beautiful objects, and they provide investigators with quite exceptional opportunities to generate results that are visually convincing and aesthetically appealing.

LBCs have been the subject of several reviews over the past ten years (Sommerville, 1977; Macgregor, 1977; Espinoza, 1980 [written in Spanish]; Macgregor, 1980; Callan, 1982; Gall et al., 1983). Current techniques for isolating, manipulating, and experimenting on LBCs have recently been described in some detail by Macgregor and Varley (1983).

In what follows, there will be frequent reference to observations and experiments on the LBCs of newts and salamanders. However, it is important to understand at the outset that LBCs are present in oocytes throughout the animal kingdom, in both invertebrates and chordates, and including the lower orders of insects. They also are present in the large primary nucleus of the unicellular alga *Acetabularia mediterranea*, and their occurrence in plants may be more widespread than we yet know. Lampbrush studies have concentrated on newts and salamanders for three good reasons. First, these animals have large genomes, and therefore large chromosomes. Second, it is easy to obtain large numbers of oocytes with good lampbrush chromosomes from newts. Third, the germinal vesicles and lampbrush chromosomes of newts are easier to handle and isolate than those of any other organism.

A set of LBCs is essentially a set of diplotene bivalents attached to one another at chiasmata. There are frequently other points of attachment between homologous and occasionally nonhomologous structures, but these usually involve only ribonucleoprotein and are not points of genetic exchange. In the LBCs of some urodeles there are substantial blocks of condensed heterochromatin flanking the centromeres, and in these species half bivalents are frequently joined by fusion of the DNP of these blocks.

LBCs vary enormously in length from one species to another according to C-value (amount of DNA per haploid chromosome set). For example, the total length of a haploid set of LBCs from *Plethodon vehiculum* ($C = 37$ pg) may be up to 20 mm (Vlad and Macgregor, 1975), whereas that of *Bipes biporus* (*Amphisbaenia*) ($C = 2$ pg) is probably never more than about 3 mm (Macgregor and Klosterman, 1979).

A single lampbrush bivalent, when examined unfixed and soon after isolation from its germinal vesicle, appears in phase contrast as a long, relaxed threadlike structure consisting of closely spaced granules—the chromomeres—connected by an exceedingly thin and almost invisible axial strand—the interchromomeric fibril (Figs. 5-1 and 5-2). Each chromomere bears one or more pairs of loops that project laterally from the chromosome axis. The average length of the loops is roughly related to C-value. The loops of the LBCs of some urodeles with large genomes (20-60 pg) extend 20-30 μm from their parent chromomeres. Those of anurans and reptiles, with genomes of between 1 and 5 pg, rarely extend more than 5-10 μm. Among the smallest LBCs that have been described are some from the oocytes of *Sepia officinalis* (Callan, 1957), the smallest microchromosomes of *Bipes biporus* (Macgregor and Klosterman, 1979), each of which probably has fewer than 10 loop pairs, and the smaller chromosomes described in the primary nucleus of the unicellular green alga *Acetabularia mediterranea* (Spring et al., 1975), where only a few of the lateral loops extend more than 2-3 μm from the chromosome axis. The overall range of variation in chromosome and loop length from one organism to another may be more than tenfold. It is important to recognize that all LBCs, from the largest to the very

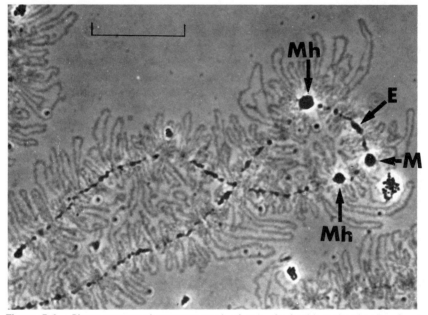

Figure 5-1. Phase contrast photomicrograph of part of a freshly isolated and unfixed lampbrush chromosome from a plethodontid salamander, showing the chromomeric axis and loops with variable lengths but similar fine scale morphologies. The chromosome was isolated in 0.1 M buffered saline, allowed to spread over the bottom of a lampbrush observation chamber, and then centrifuged to flatten the axis and loops into one plane (see Macgregor and Varley, 1983). E, the fused ends of the 2 half bivalents; M, a marker structure attached to the chromosome axis near the end of one half bivalent; Mh, a pair of marker structures located at homologous positions on each half bivalent. Scale bar = 30 μm. (Reproduced with the kind permission of Dr. J. Kezer, University of Oregon.)

smallest, have the same detailed form—chromomeres, an interchromomeric fibril, and pairs of lateral loops. That this form is so consistent over such a wide scale must be significant in relation to the manner in which DNA sequences are represented and arranged in large and small genomes, and the ways in which these sequences are packaged and expressed.

Diplotene half bivalents may of course be expected to be double structures consisting of two chromatids that are products of the last premeiotic

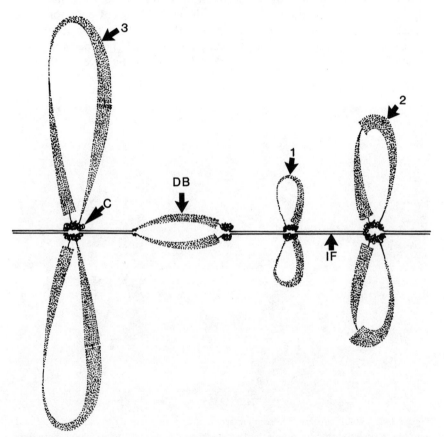

Figure 5-2. Diagram representing the main features of the currently accepted model for lampbrush structure. *IF*, interchromomeric fibril consisting of two closely apposed chromatid strands; *C* chromomere; *DB*, a pair of loops that have been drawn out into a double bridge: the drawing shows most of the chromomeric material at the end of the double bridge that follows the thick ends of the loops. 1, a pair of loops each consisting of one unit of polarization (transcription unit); 2, a pair of loops each consisting of two transcription units with the same direction of polarization; 3, a pair of loops each having three transcription units with opposite polarities.

S-phase. Reconciliation of this fact with lampbrush structure as seen with a phase contrast microscope was one of the major steps forward in lampbrushology, and it came about through critical microscopy and sheer ingenuity of interpretation by H. G. Callan when he was investigating the LBCs of crested newts in the 1950s. Callan (see Callan, 1982) rightly attached significance to three important observations. First, he knew that loops were always paired, and that "sister" loops were always alike in length and fine scale morphology. Second, he observed that, when the chromosomes were stretched to breaking point, the first breaks occurred alongside or transversely across chromomeres between the points at which the loops were inserted such that the loops were left spanning the gap and maintaining the linear integrity of the chromosome. It was mainly on account of these "double bridge breaks" (Fig. 5-2) that Callan came to associate the twoness of loops with the doubleness of meiotic chromosomes. He was particularly impressed by the fact that some loop pairs *always* exist as double bridges, implying that the linear integrity of intact LBCs still inside the germinal vesicle may depend on loops. He therefore thought that loops should have an axis of some kind that represented part of the main structural thread of the chromosome. The loop axis was demonstrated by electron microscopy in 1956, by Gall. Two years later it was shown to consist of DNA (Callan and Macgregor, 1958). Third, Callan noticed that there were certain regions of the LBCs of crested newts where the main chromosome axis was always double (Fig. 5-3), with sister chromomeres on both strands. When loop bridges were formed in these regions each was spanned by a single loop, not a pair. It was from these simple but painstaking observations that the significance of the gross structure of LBCs was correctly judged, and the model shown in Figure 5-2 was confirmed.

There are thousands of loops in a set of lampbrush chromosomes from a newt or salamander. The vast majority of these loops are more or less alike in their fine scale morphology, and it is not usually possible to distinguish one loop from the next or to identify a specific loop in several preparations from the same animal. However, loops with quite distinctive morphologies regularly occur at particular loci on the chromosomes and these *are* easily distinguishable from one preparation to another. The positions and appearances of such "marker" loops are constant and dependable at the subspecies level. Marker loops have been used, together with chromosome relative lengths and centromere indices, to construct working "maps" of the LBCs of several amphibian species that are particularly favored for lampbrush studies (Callan and Lloyd, 1975). Besides the distribution of marker loops, LBCs show other highly species or race-specific characteristics that should not be overlooked when considering the biology of these objects. The chromosomes of two subspecies of crested newt, for example, are easily distinguish-

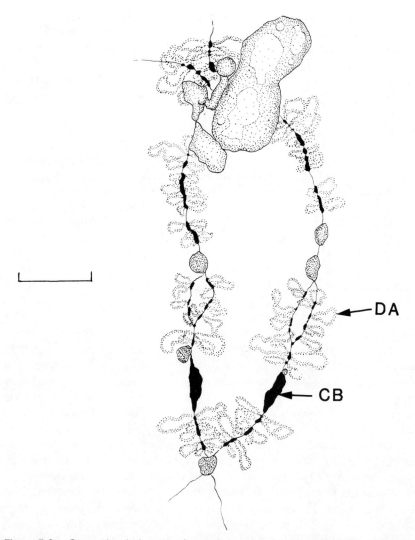

Figure 5-3. Camera lucida drawing of part of a lampbrush bivalent from *Triturus cristatus karelinii*. CB, dense bars of compacted loopless heterochromatin surrounding the centromeres; DA, a region of the chromosomes in which the axis is double, with each chromomere carrying only a single loop. A large marker structure, the "giant fusing loop" (Callan and Lloyd, 1960) is shown at the top of the diagram. Scale bar = 10 μm.

able at corresponding stages of development without any regard to markers or chromosome lengths. The LBCs of *Triturus cristatus cristatus* are generally exceedingly slender and delicate objects with small discrete chromomeres, and it is difficult to isolate them manually without causing breaks. Those of *T.c.carnifex*, on the other hand, have much stronger axes, slightly larger chromomeres, and are more resistant to damage during isolation procedures. It may be significant that the C-value for *T.c.carnifex* is 23 pg, whereas that for *T.c.cristatus* is 19 pg (Olmo, 1983). The LBCs of *T.c.karelinii* are even more robust and chromomeric in appearance. Its C-value is not yet known, but it may be reasonable to anticipate that it will prove to be greater than either of the other two subspecies.

Early in their studies of the LBCs of newts, Callan and Lloyd (1960) noted that certain marker loops could be either "homozygous" or "heterozygous" with respect to their presence or absence in all oocytes from a single animal. They concentrated on three pairs of marker loops in *T.c.carnifex* and determined the frequency of heterozygous and homozygous combinations in 22 individuals from one population of newts. They calculated the Hardy-Weinberg figures of expectation for these combinations, and found them to be in accordance with the frequencies actually observed. Accordingly, they argued that the capacity to form particular loops is a direct expression of local genetic constitution. Following these observations there was a repeated tendency to assign some sort of genetic unity to a loop/chromomere complex. But as we now know, and will see later in this chapter, the situation is not so simple. Recent studies with in situ nucleic acid hybridization have shown that heterozygosities for presence or absence of particular loops are common (Old et al., 1977; Macgregor, unpublished observations), and they are a matter that should be examined and evaluated very cautiously indeed.

Even after many years of study, it is still not known how LBCs are initially formed. In newts and salamanders and in some anurans they can be isolated in reasonably good condition from all vitellogenic and large previtellogenic oocytes. In smaller oocytes, the GV is difficult to handle and it is virtually impossible to obtain intact, well spread preparations of the chromosomes using normal manual techniques. There are only a handful of clues as to what happens to the chromosomes between the end of pachytene and the appearance of truly chromomeric and loopy LBCs in small previtellogenic cells.

It has been shown by spreading the contents of small GVs and examining them in an electron microscope that in *X. laevis* oocytes of less than 100 μm diameter there are transcription units that might reasonably be expected to have been derived from lampbrush loops, and there is a great deal of nontranscribed nucleosomal chromatin (Hill and Macgregor, 1980). So LBCs are probably beginning to take shape in oocytes of around 50 μm diameter, which is the start of Dumont stage 1 (previtellogenic, 50-300 μm diameter). The best evidence comes from thick sections of very small (between 30 and 100 μm diameter) newt oocytes, stained with haematoxylin or with Feulgen/light green (Callan, pers. commun.). In pachytene nuclei (20-25 μm diameter) the chromosomes are whiskery strands with irregular and ill-defined axes and no discrete chromomeres. The onset of diplotene in nuclei of about 30 μm diameter is clearly defined and marked by light green staining around chromosome axes that have suddenly become compact and smooth in outline. It is at about this stage that particles of amplified ribosomal DNA have dispersed over the inner surface of the nuclear envelope, but the extrachromosomal nucleoli have not yet formed (Macgregor, 1972). Discontinuities in the chromosome axes, that is, the first signs of chromomeres, are evident when the nuclei have reached 60-70 μm diameter, but discrete chromomeres comparable in size to those seen in isolated lampbrush chromosomes only become apparent in nuclei of about 100 μm diameter.

A general impression of the early changes in chromosome morphology can be gained from Figures 5-4 to 5-6. There is then a gap in our knowledge until the oocytes reach about 300 μm diameter, which is the smallest size in which true LBCs have been observed (Callan and Lloyd, 1960). In the multinucleate oocytes of *Flectonotus pygmaeus* (Anura) (more than 2000 nuclei per oocyte) the chromosomes of most nuclei are short and slightly fuzzy diplotene bivalents. It is only in those nuclei that come to lie around the outside of the nuclear cluster and swell up to look like typical GVs that the chromosomes become obviously lampbrushy (Macgregor and del Pino, 1982). It may be significant that in *Flectonotus*, as well as several other organisms that have been examined, the emergence of a distinct lampbrush form coincides with the appearance and growth of multiple oocyte nucleoli that are the products of amplified extrachromosomal ribosomal DNA (see Macgregor, 1972). Little more can usefully be said at the moment about lampbrush formation, but it is a matter that needs more investigation. For that to succeed, some new technologies and approaches will have to be invented.

At the end of oogenesis the process goes into reverse. The chromosomes contract and their loops become shorter and fewer in number. It all happens rather gradually at first, beginning in oocytes of about 1.2 mm diameter in *Triturus*; and then, quite suddenly as the oocyte approaches maturation, the chromosomes condense to less than one tenth of their full lampbrush length, all the loops disappear, and chromosomes and nucleoli become packed into a tight ball in the center of an enormous translucent GV (Callan, 1966). The end of the lampbrush phase is easy to study, and it is surprising that it has not

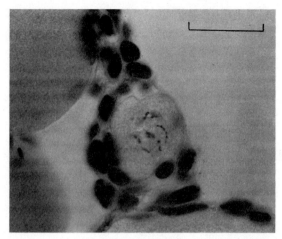

Figure 5.4. Haematoxylin stained section through a late pachytene *Triturus* oocyte showing whiskery chromosomes with irregular and poorly defined axes. Scale bar = 10 μm.

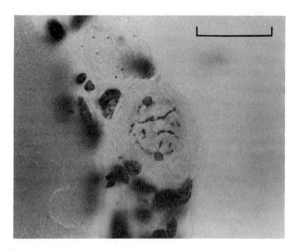

Figure 5.5. Haematoxylin stained section through an early diplotene *Triturus* oocyte with chromosomes having more compact and smooth outlines. Scale bar = 10 μm.

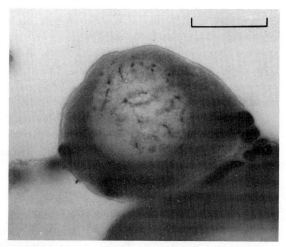

Figure 5-6. Haematoxylin stained section through a later diplotene oocyte in which the chromosomes show signs of a chromomeric organization and have some lateral fuzzy material that may represent the first stages of loop development. Scale bar = 10 μm.

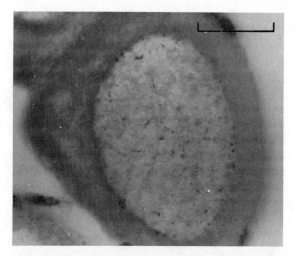

Figure 5-7. Haematoxylin stained section through a larger diplotene oocyte in which the chromosomes are diffuse and indistinct and have probably become finely chromomeric with many small slender loops. Beyond this stage, as the nucleus enlarges, it becomes difficult to distinguish loops or continuous runs of chromomeres except in thick sections. Scale bar = 10 μm.

aroused more curiosity, particularly with regard to modes of transition from active to inactive chromatin.

THE COMPONENTS OF LBCS

The Interchromomeric Fibril

The interchromomeric fibril was first visualized by Callan and Tomlin (1951) in electron micrographs of whole mounts of spread LBCs. It appeared as a single fiber, just as it subsequently did in the light microscope and has done in all but one of several studies by electron microscopy. Only Bakken and Graves (1975), who examined LBCs in a scanning electron microscope, claim to have resolved the two chromatids of a lampbrush half bivalent in a region between chromomeres. Mott and Callan (1975) have published one particularly handsome electron micrograph of a longitudinal section through the axis of a newt LBC. It shows two significant features. First, the interchromomeric fibril appears as a single straight thread estimated to be a little more than 10 nm in thickness. Second, there is a sharp transition between the fibril and the chromomeres that it connects, which are essentially two large compact roundish chromomeres, each a little less than a micron in diameter, linked by a very thin strand.

Two questions would seem to be worth asking about this fibril. Does its chromatin have the same nucleosomal organization as that of the chromomere, and is it ever transcribed? Neither question has yet been tackled.

The Chromomere

The chromomeres of fully extended LBCs appear in phase contrast as compact refractile granules, sometimes obviously single and rounded and bearing just one pair of loops, but often compound with several pairs of loops emerging from an elongated irregularly shaped object. In most urodele species that have been examined, single chromomeres are a little less than 1 μm in diameter, and are spaced out at intervals of about 2 μm between centers (Vlad and Macgregor, 1975). They stain strongly with Giemsa, hematoxylin, and Feulgen. There is therefore no doubt that they are real DNA-rich structures, and signify an alteration in the form of the axial chromatin between chromomere and interchromomeric fibril.

In general, there is no convincingly consistent pattern of chromomere distribution along a particular chromosome arm, but if the chromomeres are carefully mapped on homologous regions of the same bivalent, then a general similarity is apparent. A study of this kind has been done on the longest chromosome (chromosome 1) of crested and marbled newts (Sims et al., 1984) and three matters are of significance. First, the number of individually distinguishable chromomeres, single or compound, is virtually the same for the short arms of both half bivalents. Second, the short euchromatic tip of the long arm of chromosome 1, represented by the last 20 μm of the LBC, always has about 20 exceedingly small single chromomeres of uniform size carrying short thin loops of uniform size and appearance. Third, there are fewer distinct chromomeres per unit of length in the long heterochromatic region of the long arm of chromosome 1 than there are in the short euchromatic arm of the same chromosome. The general impression is that the lampbrush chromomeres seem larger and more often locally clumped in regions that are heterochromatic in mitotic chromosomes (Giemsa C-positive) than in other places. An extreme example of correspondence between Giemsa C-band heterochromatin and lampbrush axial structure is found around the centromeres of the newt *T.cristatus karelinii*. In this animal all but one of the chromosomes have massive blocks of pericentric heterochromatin (Sims et al., 1984). These consist largely of two highly repeated short DNA sequences (Baldwin and Macgregor, 1985). In the LBCs of this animal there are conspicuous bars of compact loopless chromatin surrounding the centromere granules (Figs. 5-3, 5-8).

Since chromomeres, whether single or compound, are optically distinct structures, it is possible to count them and thereby determine the total number for a complete haploid set of LBCs. Chromomere counts have been made for fully extended LBCs of certain plethodontid salamanders (Vlad and Macgregor, 1975). The outcome of that study was simple, clear, and important in relation to any discussion of the genetic nature of a lampbrush chromomere. The animals used to belong to the same genus, and are remarkably alike in size and appearance even though they have been separated as species for at least 80 million years. They have the same chromosome number, and their chromosomes have the same relative lengths and centromere indices. However, their C-values differ by a factor of nearly 2:1. The chromosomes of one species are therefore much larger than those of the other. The species with the larger genome has at least 60% more chromomeres than the one with the smaller genome. There can be no shadow of doubt that these counts are meaningful and significant. Chromomere size and distribution are obviously the same for both species, but in one species the LBCs are much longer *at any stage* than in the other. These observations dispelled any notion that lampbrush chromomeres should be equated with genes in the sense that has applied to the bands of polytene chromosomes. They are certainly units of some kind, and since they carry transcription units in the form of loops, it seems likely that they bear some relationship to the arrangement of expressible DNA sequences along the chromosomes. The nature of this relationship, though, remains obscure.

The internal structure of chromomeres, as seen in electron microscope sections, reveals nothing other than that they are fibrous aggregates (Mott and Callan, 1975). Miller spreads of LBCs show vast amounts of nucleosomal chromatin that most likely represents dispersed chromomeric material (Hill, 1979). Several small but nonetheless very important questions remain. Where are the bases of the loops inserted in relation to the body of a chromomere? Does a loop begin "outside" or at the edge of a chromomere, and end somewhere in its main body? What happens to a chromomere when a double bridge break happens? Does it split into two parts, or do the beginnings of the loops follow directly from the adjoining stretch of interchromomeric fibril, leaving their ends attached to all or most of the chromomere? In functional terms, is there any relationship between chromomeres and the points at which transcription begins on loops? Is a chromomere just a "ball of wool" from which a particular set of DNA sequences is pulled for transcription, or is it a mass of condensed chromatin that immediately follows downstream from a specific set of transcribed sequences?

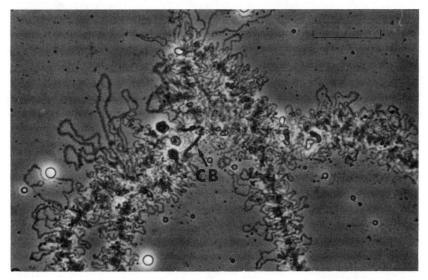

Figure 5-8. Phase contrast micrograph of part of a lampbrush bivalent from *T.c.karelinii*, showing the compact bars of loopless heterochromatin (CB) that surround the centromeres of all lampbrush chromosomes in this sub-species. Scale bar = 20 μm.

The Centromeres

Very little is known about lampbrush centromeres, yet they do represent a rare instance in which a centromere is clearly resolvable as a very small, round, positively staining granule. One might therefore suppose that if centromeres had structure or distinctive molecular features then centromeres of LBCs would be good for investigation. However, all we know is that the centromere granule that we see on a lampbrush chromosome stains with Feulgen and therefore contains DNA. It carries no loops, stains differently from chromomeres with Giemsa, has a distinctive surface structure when examined with an SEM (Macgregor and Swan, unpublished observations), and occasionally fuses with its partner on the other half bivalent. Nothing more can usefully be said.

The Loops

In its simplest form, a lampbrush loop consists of an axis of DNA with RNA polymerase molecules distributed along it at a density of about $20/\mu$m. A nascent RNA transcript is attached to each polymerase. The transcripts increase in length from one end of the loop to the other, such that the last and longest transcript should have the same length as the DNA axis from which it was transcribed. RNA transcripts are complexed with proteins as they form, and are packaged into ribonucleoprotein particles that make up much of the fine texture of the loop as seen with phase contrast. Most loops start with a thickness approaching the limits of resolution of the light microscope and are perhaps 2-3 μm wide at their thick ends. These dimensions apply to loops ranging in length from 10-50 μm or more. A detailed account and discussion of loop structure has been given in a previous review (Macgregor, 1980). The account offered here will specifically concentrate on those characteristics of loops that help toward an understanding of their genetic significance and their transcriptional role in the GV.

A simple loop that progresses from thin to thick along its entire length (Fig. 5-9) represents a unit of transcription (TU) in the sense that RNA polymerase molecules attach to a recognition site at the thin end of the loop and move around the loop axis to the thick end, transcribing RNA as they go. There are, however, several important variations on this basic plan.

Loops may consist of several TUs arranged in tandem, so that the matrix of one loop may show several gradients of thin to thick arranged with the same or opposite polarities (Figs. 5-2, 5-10). The different arrangements of TUs that have been identified were classified by Scheer et al. (1976). In such cases each TU is generally separated from the next by a small segment of untranscribed chromatin (Angelier and Lacroix, 1975; Scheer et al., 1976, 1979).

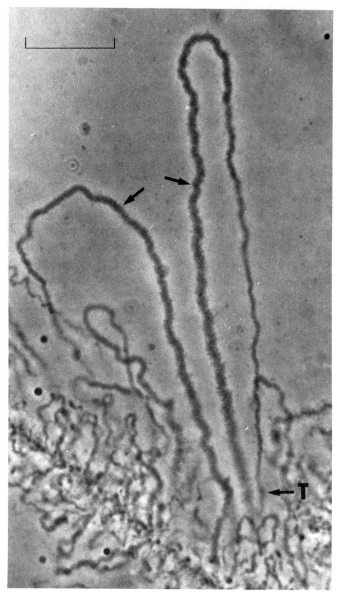

Figure 5-9. Phase contrast micrograph of part of a lampbrush chromosome showing a pair of long lateral loops (arrows) each consisting of a single unit of polarization (TU). *T*, the thin end of one loop of the pair. Scale bar = 20 μm.

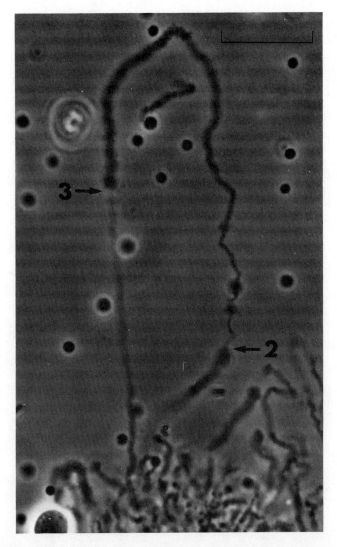

Figure 5-10. Phase contrast micrograph of a long lateral loop that consists of three sequential TUs each with the same direction of polarization. The numbers 2 and 3 indicate the starts of the second and third TUs. Scale bar = 20 μm.

Certain loop complexes may exist either as single matrix gradients or as longer loops with multiple matrix units all of the same polarity. The large loops near the centromere of chromosome 2 in *Notophthalmus viridescens* are a well documented example (Gall, 1954; Gould et al., 1976; Hartley and Callan, 1978; Callan, 1982).

Scheer et al. (1979) have described single TUs where the transcripts become progressively longer up to a point, and then quite abruptly appear shorter, producing a visible "step-down" in the unit's polarization.

The spacing of polymerases along a loop axis may be quite variable. Hill (1979) observed TUs of up to 10 μm in length (equivalent to at least 30 kb of DNA) in *X.laevis* oocytes, and reported variation in the spacing between polymerases over at least an order of magnitude.

These facts tell us several important things about loops. First, an average lampbrush loop represents a lot of DNA, certainly much more than would be needed for the transcription of most simple proteins. Second, a loop need not be regarded as a unit of any kind, although we refer to it as a transcription unit when it shows just a single matrix gradient. Third, those TUs that show a step-down in polarization signify that there may be some excision or "processing" of transcripts while they are still being transcribed from the loop axis. Last, if polymerase packing density on the loop axis is related to intensity of transcription, then different loops may transcribe at different rates. It should be said that although close packing of polymerases has always been presumed to indicate a high level of transcription, this need not necessarily be so. Very little is known about actual rates of transcription in relation to polymerase packing density, and it seems possible that densely packed TUs may reflect a "traffic jam" situation in which the polymerases, although abundant, are relatively slow moving.

Three more matters need emphasizing with regard to the form and behavior of lampbrush loops. First, the two loops that emerge from a loop pair/chromomere complex, commonly referred to as "sister" loops, are almost always identical in length, fine scale morphology and polarity of TUs. This is indeed a remarkable feature of LBCs, and more will be said of it later. Very rarely, sister loops are of different lengths. Indeed there is only one example of nonidentical sisters (Morgan et al., 1979; Macgregor, 1980), although others have been seen from time to time. Nevertheless, the fact that they exist at all must enter into any consideration of the molecular basis of loop structure and function.

Second, there is definitely *not* a one-to-one ratio between loop pairs and chromomeres. So counts of chromomeres are not a reliable measure of the number of loops. Even more so, counts of chromomeres tell us nothing about the number of TUs, since many loops include more than one TU.

Third, it is certain that some loops are present throughout the greater part of oogenesis, and are always found in the same positions on the chromosome set. However, it is important to remember that we are referring here to loops with morphologies that are distinctive enough for them to be easily recognized. These are the loops that form the basis of the working maps of LBCs. We do not know whether the same applies to any of the great number of nondescript loops that cover most of the chromosome set.

At this point it seems appropriate to introduce what is perhaps the most fundamental unanswered question about LBCs. Is the lampbrush loop a permanent structure in the sense that it is established as a TU at the start of oogenesis and remains as such with the same point of initiation through to the end of the lampbrush phase? If a loop is a permanent structure, then within a particular chromosome segment or even a particular chromomere is it always the same piece of DNA that forms the loop in every oocyte from every animal and at all stages? Or is the loop a transitory structure in the sense that its transcriptive activity forms part of a complex program of gene activity during oogenesis, certain DNA sequences being exposed for transcription in loop form at one stage and being replaced by others later on? In this context, one small matter should be kept in mind. If loops can form at a rate approximating to the generally estimated speed of movement of RNA polymerase along a DNA strand, then we might expect them to be quite dynamic structures with the potential to emerge and disappear in a matter of hours. To be sure, those loops and other objects that have been used for the construction of LBC maps are not transitory, but there are very few of them in relation to the total number of loops in an entire set of LBCs and their distinctive forms suggest that they may represent something special.

Two more questions remain, and neither can be answered directly. How much of the chromosomal DNA is represented by the axes of the loops? Is all the transcriptive activity of LBCs confined to visible loops? The first of these questions has been discussed previously in some detail (Macgregor, 1980). If there are 5000 loops, each about 30-50 μm long and with their DNA in the B form, then in a genome of 30 pg all of the extended loop DNA will account for less than 2% of that genome. However, that is almost certainly an underestimate on several counts. The figure of 5000 is based on chromomere counts, so there may be as many as twice that number of loops. The DNA of the loop axis is most likely not all in the extended B form. Regions between polymerases may be complexed into nucleosomes, and a higher order of secondary structure seems likely in many of the larger and more conspicuous loops. And as has been mentioned already, we cannot be sure that loops are permanent structures so that several different parts of the DNA in one chromomere may be extended at different times during the lampbrush phase. It therefore seems reasonable to expect that anything up to 10% of the genome is involved in loop formation during the lampbrush phase.

Whether transcription is confined to visible loops is hard to say. Neither ^{3}H uridine incorporation nor DNA/RNA-transcript in situ hybridization have provided convincing evidence of transcription on chromomeres or on the interchromomeric fibril. The meager evidence on this topic comes from three quite different approaches. First, DNA/RNA-transcript in situ hybridization of an abundant centromerically localized satellite DNA to the LBCs of *T.c.karelinii* reveals the presence of very small amounts of satellite transcript on and around the heterochromatic bars that flank the centromere granules (Macgregor and Baldwin, 1985). No loops are visible, so we must suppose that transcription is happening on the surface or at the extreme ends of the heterochromatic bars. Second, in scanning electron micrographs of LBCs the components of the chromosome axis, when they can be distinguished among a dense forest of loop material, have the same general kind of surface structure as other compact nonaxial ribonucleoprotein objects associated with the chromosomes; they definitely do not look as one might expect if they were chromatin of the kind found in mitotic chromosomes (Harrison et al., 1983). This is particularly so in the case of *T.c.karelinii* centromere bars. These are known to be compact loopless chromatin, yet they have a surface structure that is quite unlike any other compact chromatin that has been visualized in the SEM (Macgregor and Swan, unpublished observations; see Fig. 5-11). Third, Scheer (1982) has described transcriptionally active chromatin from oocytes of *Plurodeles*, which is not in the form of polarized units, but appears as a periodic alternation of thick and thin regions occurring in clusters of around 10,000 repeats. Whatever this chromatin represents, it is significant insofar as it is transcriptionally active and it is of a form that would not be expected to produce a visible polarized loop.

THE PROTEINS ON LBCS

We see LBCs with the light microscope largely on account of their proteins and the effect they have in transforming nascent RNA transcripts into discrete ribonucleoprotein particles. A considerable amount is now known about these proteins, and much of the available information has been brought together by Sommerville in his 1981 review, from which most references in this section have been drawn. Only the basic essentials of the situation will be mentioned here, together with some observations that clearly relate certain proteins in the GV to transcriptive activity and the general behavior and appearance of the chromosomes.

In the first place, there are the histones. These present quite a simple situation insofar as they are present in the nucleosomal chromatin of the chromosome axis (chromomeres and interchromomeric fibril) and are also present at much lower concentrations in the loops. The presence of histones in the loops is thought to result from adsorption of some of the abundant histone of the GV-plasm onto the negatively charged components of the loop matrix.

With regard to other proteins, the situation is less well understood, but there is some useful information available. On the one hand, there would seem to be a small number of polypeptides (MW 34,000-40,000) that are largely responsible for generating and maintaining the particulate structure of the nuclear RNP, and another small group of polypeptides (MW 50,000-68,000) that seem to be intimately associated with the primary transcripts of lampbrush loops. However, having said this, it is important to add that the range of polypeptides that are recoverable from the GV is very wide indeed, although the function and distribution of most of these is not clear. On the other hand, there are undoubtedly some proteins that are confined to certain specific loops. In one immunofluorescent study, for example, a particular anti-oocyte nuclear RNP reacted with only 10 pairs of loops, and two proteins that are complexed with 4S and 5S RNA in cytoplasmic 42S storage particles produced antibodies that reacted with only a few specific loops. One of these actually stained a loop that has been identified as a site of 5S RNA genes. It is of particular interest that some immunostaining reactions that label specific

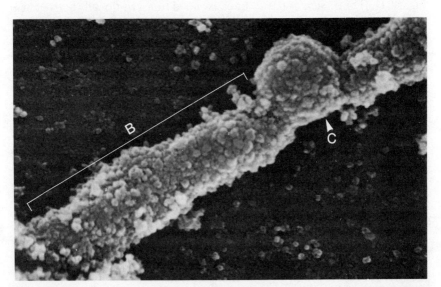

Figure 5-11. Scanning electron micrograph of part of one of the heterochromatic centromere bars on a lampbrush chromosome from *T. c.Karelinii*. The chromosome was isolated in 0.1 M buffered saline, fixed in glutaraldehyde/osmium, critical point dried and then thinly sputter coated with gold. B, heterochromatic bar; C, centromere granule.

loops stain only part of the loop, even though the RNP matrix is continuous over the whole length of the loop.

A backup to the conclusions from immunofluorescent work has been provided by AgAs silver staining method of Goodpasture and Bloom (1975). Varley and Morgan (1978) showed convincingly that several of the principal marker structures on the LBCs of *T.c.carnifex* were AgAs positive, as were about 15 more "normal" pairs of loops. Silver staining was clearly not a simple function of the particular form of a loop or the density of its matrix, since in some cases where particular loops appeared in widely different forms they were nonetheless always AgAs positive. The AgAs technique is normally associated with identification of active nucleolus organizers on chromosomes (O. J. Miller et al., 1976; D. A. Miller et al., 1976). None of the loops that showed as AgAs positive on LBCs were known sites of nucleolar genes, and there was no silver staining of loops in the chromosome region that *is* known to include the main nucleolus organizers.

All of these observations taken together suggest that while there are certain classes of polypeptide that are included in the RNP of the GV, there *is* RNA-sequence-specific binding of protein in some loops. Just how widespread are these sequence-specific proteins remains to be seen.

A different kind of interest may be attached to the results of experiments in which anti-histone, anti-HMG and anti-RNA polymerase II have been applied to LBCs, or experiments in which drugs that inhibit RNA transcription (actinomycin-D, a-amanitin) have been used on oocytes, isolated GVs or isolated LBCs (Bona et al., 1981; Scheer et al., 1979a; Kleinschmidt et al., 1983; Snow and Callan, 1969; Mancino et al., 1971; Morgan et al., 1980). In most of these cases the ultimate effect has been a dispersal of loop RNP, retraction of the loops, and contraction, often quite dramatic, of the chromosomes; and in most cases the effects are reversible, so that when the causative agent is removed from the oocyte the chromosomes re-extend and their loops reappear. None of this is entirely surprising, and it could be explained by supposing that extension of the loops is dependent upon continued occupancy of them by actively transcribing RNA polymerases and the maintenance of an appropriate balance between the state of the chromosomal DNP and its accessibility to RNA polymerase. Having drawn attention to the ease with which LBCs can be stripped and contracted by agents that stop transcription, it is very important to stress that under certain carefully controlled conditions it is possible to stop transcription, as judged by nonincorporation of tritiated uridine into loop RNA, *without* removing their matrix (Snow and Callan, 1969; Morgan et al., 1980).

More problematical are the effects of anti-actin and actin-binding proteins (Scheer et al., 1984) and the effects of X-rays (Pellatt and Macgregor, unpublished observations). All of these cause loss of loop RNP and dramatic contraction of the chromosomes to less than one-tenth of their original lengths. In the case of X-rays the effects are quite remarkable. The LBCs of small vitellogenic and previtellogenic oocytes are unaffected. The chromosomes of larger vitellogenic oocytes (more than 1.2 mm diameter in *T.c.carnifex*) show complete contraction within 1-2 days of irradiation (whole body irradiation of the newt with 1500 rads). About seven days after irradiation the chromomosomes begin to re-extend, and their loops reappear. Within 14 days they are back to normal. The meaning of these events is at present obscure, but the system is undoubtedly promising for the study of mechanisms of loop formation, among other things. Perhaps most exciting of all is the question: When the chromosomes recover from the effects of X-rays, are precisely the same pieces of DNA used to form the new loops?

EXPRESSION OF DNA SEQUENCES ON LBCS

In a set of LBCs from a European or American newt the average length of the loops is around 50 μm. representing at least 150 kb of DNA. The average lengths of the TUs may be shorter since there is often more than one TU per loop. Loops with more than three TUs are rare. There are at least 4,000 chromomeres per haploid set of LBCs. Allowing for an average of 1.5 loop pairs per chromomere and two TUs per loop, there are therefore around 12,000 TUs per chromosome set.

The DNA axis of virtually all lampbrush loops contains substantial amounts of repetitive DNA, some of it highly repetitive satellite DNA, but most of it belonging to the "middle repetitive" class (Barsacchi and Gall, 1972; Sommerville, 1977; Varley et al., 1980; Macgregor, 1980). We suppose that the majority of this DNA is noncoding, yet it is liberally transcribed on the long TUs of LBCs. One such case has been investigated in detail. This histone gene cluster in *N.viridescens* seems to occupy the thin ends of several loops, and is followed by runs of highly repeated DNA in the remainder of the "histone" loops. These runs of satellite may be up to 35 μm long (Gall et al., 1981; Diaz et al., 1981).

There appears to be no relationship between the abundance of a particular sequence, whether "functional" or not, and the level to which it is transcribed in an oocyte nucleus. There are many thousands of 5S genes (Hilder et al., 1983), but only a few of these can be transcribed (by RNA polymerase III) on lampbrush loops. There are likewise hundreds of copies of the histone gene complex concentrated at two main loci in *N.viridescens* (Gall et al., 1981; Diaz et al., 1981), yet no more than 10 or 20 of these may be transcribed on the histone loops. There are probably hundreds of thousands of copies of some tandemly repeated and transcribed satellites (Varley et al.,

1980; Jamrich et al., 1983), but some are transcribed on only 10-20 loops, equivalent to about 6,000 copies of a sequence 500 bp long.

One further point that needs re-emphasizing here is the size of an average structural gene as compared with the length of an average TU. An average gene, coding for a single protein, may be expected to contain up to 5 kb of DNA. Such a gene would make a TU of around 2 μm in length, which would scarcely be visible or recognizable in the light microscope. Even a gene cluster such as the histone complex would make a TU of only 3 μm. Of course, there are likely to be much larger gene complexes, but at present we know nothing about them in animals that are commonly used for lampbrush studies.

The current explanation of the lampbrush phenomenon is based on the supposition that the chromosomes are actively providing something for the oocyte and the developing embryo. At the thin base of the loop or a TU there is a promoter site for a "structural" gene, followed by the gene sequence itself. RNA polymerase molecules recognize this promoter and attach to it, and transcription is initiated. The polymerases then move along the loop through the structural gene, producing RNA transcripts that remain attached to the loop (TU) axis. When the polymerases reach the end of the structural gene they ignore whatever termination signals may be present in the DNA sequence and read on past the end of the gene into whatever follows downstream. In loops that have more than one TU the polymerases continue to transcribe until they approach the start of another TU that is already initiated, where they stop transcribing and the loop changes abruptly from thick to thin. We should recall that in electron micrographs of tandem TUs there is usually a short "spacer" of untranscribed DNA separating the end of one from the beginning of the other (Angelier and Lacroix, 1975; Scheer et al., 1976). In loops that consist of just one TU the polymerase continues along the loop axis until it meets DNP that is for some reason physically untranscribable. In most cases this is the chromomeric DNP of the chromosome axis.

This explanation has come to be known as the "read through" hypothesis. It was first suggested by Old et al. (1977), later supported by Varley et al. (1980) and Morgan et al. (1980), following their experiments on transcription of satellite and ribosomal DNA on newt LBCs. Hard evidence and a definitive statement of the hypothesis was then provided by Gall et al. (1981) and Diaz et al. (1981), based on studies of the transcription of the histone gene complex and associated satellite DNA on the LBCs of *N. viridescens*. It is an extremely useful hypothesis, but of course (and fortunately for lampbrushologists) it leaves several important questions open.

Why are there so many TUs? Why does the number and length of the TUs vary according to C-value, LBCs in small genomes having fewer and shorter loops than those from large genomes? Why is it that only a few members of each multigene family or array of tandemly repeated sequences are transcribed at any one time in a GV?

The answer to the first of these questions has to be that we simply do not know, and will not know until we understand better the basic significance of the lampbrush phenomenon. The answer to the second question is easier. Large genomes have more duplication of functional genes and greater diversity and copy numbers of repetitive sequences. So we might reasonably expect more functional gene promoters and greater lengths of noncoding repetitive DNA between them, leading to more and longer loops. It should be stressed that the latter is a *suggested* answer, not a definitive one.

The answer to the third question is altogether more challenging and interesting. The question itself leads us to focus on two possibilities. The first is that loops are permanent structures in the sense that the piece of DNA that forms the loop axis and the promoter at its base are the same throughout the greater part of the lampbrush phase. The second is that each loop/chromomere, whether it be complex or uniformly repetitive in sequence content, may contain many potential promoters, and some of these, but not necessarily always the same ones, are selected or enabled to form TUs. In the first case, the arrangement of loops on LBCs within a particular species should always be the same, allowing for genuine heterozygosities, in the second case it should rarely be the same except for loop-chromomere complexes that consist almost entirely of tandem repeats where we would not expect to be able to distinguish by any means between expression of one group of repeats and another. Reduced to practical terms, the problem is straightforward and should not be hard to resolve. It is well known that certain DNA sequences that have been used as "probes" for in situ hybridization experiments bind only to certain parts of some loops when hybridized to the loop RNA. Where one can identify a loop that shows a partial and clearly identifiable pattern of labeling, does the sister loop show exactly the same pattern? Is there a loop with exactly the same pattern of labeling at the corresponding position on the other half bivalent? Are there loops with the same pattern of labeling at corresponding positions on the lampbrush chromosomes of other oocytes of the same animal?

Sister loops arising from the same chromomere nearly always have the same length and appearance, and in autoradiographs of DNA/RNA in situ hybrids they show the same pattern of labeling (Old et al., 1977; Varley et al., 1980; Diaz et al., 1981). Although a few exceptions have been seen (Morgan et al., 1980), it would seem that whatever we say about LBCs we must take full account of sister loop identity. Therefore, if within a particular loop/chromomere complex the promoters for the start of transcription are always the same on the two sister loops, then how are they co-selected? Could co-selection signify something about the organization of a chromomere? Is a chromomere just like a ball of wool, or is its chromatin packaged in some specific way such that particular parts are always in the same positions with respect to the bulk of the chromomere on both halves (chromatids) of the chromosome? Chromomeres undoubtedly do have some structure, otherwise they would not break cleanly to form double bridges. Are they perhaps more organized than we have hitherto imagined?

Our own and Vlad's (unpublished) in situ hybridization experiments with ribosomal and satellite DNAs and with some heterologous probes from low copy number genes (tubulin, actin, globin) have presented us with a most confusing set of results, among which we can find little evidence of consistency with respect to the positions or labeling patterns of partially labeled loops. Something of the kind of variation that has been encountered is shown by Morgan et al. (1980), where labeled loops occupied widely variable positions on the lampbrush chromosomes 1 of 12 different newts all belonging to the same subspecies and population. In a recent study of the positions of loops showing distinctive partial labeling patterns after in situ hybridization with a mixed low Cot probe and with a probe comprising most of the ribosomal gene repeat from *X.laevis*, we have observed some examples of loop pairs with identical labeling patterns at corresponding positions on 2 half bivalents, but they have been the exception rather than the rule. Most distinctively labeled loops have been present on only one lampbrush half bivalent, and would be described in the classical sense as "heterozygous" for presence or absence. In only a few cases have we seen loops with identical labeling patterns at corresponding loci on two or more preparations from different oocytes from the same animal. However, in this context it is important to note that Old et al. (1977) and Diaz et al. (1981) did see several examples of partially labeled loops in their in situ hybrids, and they specifically state that certain of these were always present at the same loci in all oocytes from the animals they used. Both of these groups of investigators were working with probes complementary to tandemly repetitive sequences.

At the present time the basic question—whether the transcriptive activity on a loop or TU represents part of a complex and purposeful program of gene activity in oogenesis—can only really be approached successfully by concentrating on patterns of transcription of well defined single or low copy number structural genes. If certain single copy genes could be shown convincingly always to occupy the thin ends of certain loops at regular positions on the chromosomes throughout, or at specific stages of oogenesis, then arguments for an oogenetic program of gene expression involving specific selection of promoters would be strong, and the read-through hypothesis would be wholly satisfying. Until such evidence is available, we cannot do better than to argue from circumstances and from what we know about transcription of functional repetitive sequences.

The situation with regard to repetitive histone, 5S and ribosomal DNA sequences is presently as follows. Histone message is transcribed more or less continuously during the lampbrush phase, and it is very probably used for synthesis of histones both during and after oogenesis (Gall et al., 1981). 5S genes are abundantly and intensively transcribed in oocytes (Korn, 1982); their activity being geared to balance the massive production of 28S, 5.8S and 18S RNA by the amplified ribosomal DNA in hundreds of extrachromosomal nucleoli. However, it does seem that the greater part of 5S gene activity happens in very small oocytes, perhaps before the major phase of lampbrush activity (Ford, 1971). At present it is quite hard to assess the scale or significance of 5S transcription as detected by in situ hybridization of 5S DNA to 5S transcripts on LBCs. Ribosomal transcription is distinctly curious. There are hundreds of free nucleoli where there is intensive and well regulated transcription of ribosomal RNA by RNA polymerase 1 (Roeder, 1976). In some, but by no means all, animals that have LBCs, there is evidence of transcription of ribosomal genes at the lampbrush nucleolus organizers. *Axolotl* (Callan, 1966), *N. viridescens* (Gall, 1954) and *Plethodon* (Kezer and Macgregor, 1973) are good examples. However, in the European crested newt, which is one of the most widely used sources of lampbrush material, the main nucleolus organizers are not expressed at all in the oocyte (Morgan et al., 1980). Besides all this, there is a small class of scattered ribosomal DNA sequences that are transcribed in the form of lampbrush loops, but not by RNA polymerase 1 (Morgan et al., 1980).

Circumstantial evidence for purposeful gene activity in oogenesis is overwhelmly strong, but it is nonetheless exceedingly hard to relate it to what we see on LBCs. There can be no doubt whatever that maternal mRNAs and specific proteins that are essential for early embryonic development are produced at some stage following pachytene, and many of these become distinctly localized in the cytoplasm of the unfertilized egg. The references on this topic are too numerous to be cited here but they relate to organisms as diverse as molluscs, echinoderms, insects, ascidians and amphibians.

Polyadenylated mRNA is synthesized in substantial amounts and diversity during oogenesis. According to Perlman and Rosbash (1978) a mature *Xenopus* oocyte may contain as many as 20,000 different poly(A)mRNA types. The same general picture emerges from the studies of Thomas et al. (1981) and Anderson et al. (1982), who draw an interesting distinction between two classes of poly(A) mRNA. Thomas et al. state that each structural gene is represented in the egg cytoplasm by several overlapping transcript forms. Some of these are mature messages which are assembled into functional polysomes following fertilization and represent the classical "maternal message" class. Others are incompletely processed transcripts that include the same message sequences found in maternal mRNA but also contain other covalently linked repeat or intervening sequence elements. Anderson et al. (1982), while agreeing in principle with previous investigators, differ in detail in so far as they find that *most* of the maternal poly(A)mRNA sequences present in the mature egg contain interspersed transcripts of repetitive DNA, and all of this RNA is unstable and depends on continual replacement from the nucleus in order to maintain its amount and sequence diversity. A useful and concise review of maternal mRNA in amphibian oocytes is given by Shiokawa (1983).

The problem is that the evidence from biochemical studies of mRNAs and clearly functional repeat sequences simply cannot be linked meaningfully to lampbrush activity. It is significant in this context that virtually all lampbrushologists have studiously avoided discussion of the role of LBCs in making maternal factors, and with few exceptions (Anderson et al., 1982: see last section of this chapter) biochemists have been just as wary of relating their findings to the cytological view of oogenesis. It seems safe only to say that during the lampbrush phase there is widespread and perhaps indiscriminate transcription on lampbrush loops of a range of highly and moderately repetitive DNA sequences, widespread synthesis of many kinds of mRNA often with associated transcripts of repetitive DNA, continuous (if modest) transcription of histone and 5S RNAs, and variable, nonexistent, or incidental read-through transcription of chromosomal ribosomal DNA. It is a messy picture, but to some extent it all makes sense if we accept that the oocyte and the developing embryo require a lot of histone, a lot of 5S RNA, and ribosomal RNA, and a wide range of mRNAs, and if we can accept that in order to accomplish these objectives it is inevitable that a lot of more or less useless repetitive DNA is transcribed at the same time.

THE SIGNIFICANCE OF LAMPBRUSH FORMATION

Over the past 20 years several hypotheses have been offered to explain why chromosomes should assume a lampbrush form. The "classical" hypothesis says that LBCs are a cytological manifestation of a system evolved for the production of many kinds of RNA that are needed for the maintenance and programing of early embryonic development. The notion was first introduced by Duryee in 1950 and later by Gall (1955), and it remains valid and interesting in principle.

The first "genetic" hypothesis of lampbrush function was the famous

Master: Slave hypothesis introduced by Callan in 1967. It drew attention to evidence that suggested that "... each unit of information encoded as a DNA base sequence is serially repeated" and "... among such serially repeated sequences a terminal unit serves as a 'master' sequence, within which recombinational events can occur, followed by 'slave' sequences that are not directly involved in recombination but that are made congruent to the master sequence once in a life cycle." (Callan, 1967). The formation of lampbrush loops was said to be the outcome of a master: slave matching process. The hypothesis was immensely attractive, and although it is no longer tenable, it should not be forgotten, if only because it represents the kind of unfettered lateral thinking that is so essential when searching for a true understanding of complex biological phenomena.

It is probably true to say that most developmental and molecular biologists now accept the classical hypothesis of lampbrush function, and most people who work on LBCs are directing their experiments toward a search for regulated programs of gene expression that relate to the needs of oocytes, eggs, and embryos. There are, however, a few inescapable biological facts that do not quite fit with the classical hypothesis, and no article on LBCs would be complete if these facts were not mentioned and discussed.

First, in all organisms that have LBCs, all the chromosomes assume the lampbrush form, including supernumeraries and microchromosomes (Kezer, unpublished observations on the supernumeraries of salamanders belonging to the genus *Thorius*). Of course, we cannot presume that because some chromosomes vary with regard to their presence or numbers in a population they have no useful genetic content. Nevertheless, if we are to suppose that LBCs do represent a complex and sophisticated program of gene expression, then the program must be flexible enough to accommodate the occassional intrusion or loss of quite large numbers of TUs. An alternative view of the problem raised by the lampbrushiness of supernumeraries is that *any* chromosome that finds itself in the peculiar environment of a GV will inevitably assume a lampbrush form: but to say that is to grossly devalue the notion of a regulated program.

Second, extra LBCs seem to confer no special advantage upon an oocyte. Triploid gynogenetic females of *Ambystoma jeffersonianum* have hexaploid oocytes with three times as many LBCs as the oocytes of their normal diploid kin (Macgregor and Uzzell, 1964). Oocytes of the Pacific tailed frog *Ascaphus truei* have eight and sometimes 16 GVs each, with a full set of LBCs (Macgregor and Kezer, 1970). Oocytes of the South American marsupial frog *Flectonotus pygmaeus* begin with over 2,000 complete meiotic nuclei, of which several hundred begin the lampbrush phase and at least 10 reach a stage of full lampbrush development (Macgregor and del Pino, 1982).

Third, there are some species of reptile that have large watery GVs, do not amplify their ribosomal genes, and do not form LBCs at any time in oogenesis. The South African lizard *Ichnotropic capensis* is such an animal. Its oocytes derive their ribosomes from hundreds of surrounding "pyriform" cells through direct cytoplasmic bridges (Andreucetti et al., 1978). Do these same pyriform cells provide the mRNA that would normally be produced by LBCs, and if they do then what of other reptilian species, such as *Bipes biporus* (Macgregor and Klosterman, 1979), where there are both surrounding pyriform cells and GVs with fully formed LBCs and amplified ribosomal DNA?

The message that comes across from these and other unusual oogenetic situations (see Macgregor, 1972 for review) is that while LBCs are found in the growing oocytes of animals belonging to all phylogenetic groups, the timing, programing and scaling of lampbrush activity do not seem to be altogether important.

Acetabularia mediterranea is the only known case of lampbrush formation, in the sense covered by this article, that does not happen in an oocyte. What does it signify? The organism is a unicellular alga that grows from a small zygote to a highly differentiated structure comprising a basal rhizoid, a stalk that may reach 3-4 cm in length, and complex sporulating cap. In the course of development the cell increases in volume from about $10^2 \mu m^3$ to $10^{10} \mu m^3$. At the same time the nucleus increases in volume by a factor of about 4×10^5, and the number of nuclear pores increases to more than 2 M (see Kloppstech, 1982). The nucleus contains around 25,000 copies of the genes for ribosomal RNA in *A.mediterranea* (Trendelenburg et al., 1974), and its lampbrush chromosomes are meiotic and unquestionably lampbrushy (Spring et al., 1975).

One essential difference between animal oocytes and *Acetabularia* is that, whereas in oocytes it seems that materials are being accumulated and stored for *subsequent* development, in *Acetabularia* it seems that the cell accomplishes most of its growth and differentiation *during* the lampbrush phase. A feature that is distinctly common to both oocytes and *Acetabularia* is ribosomal gene amplification and synthesis of ribosomal RNA on a massive scale. If we consider the nucleus of *Acetabularia* to be equivalent to a germinal vesicle, then it would seem appropriate to ask which of three major events leads the way to the formation of such a nucleus: a purposeful program of transcription on LBCs, amplification and transcription of ribosomal genes, or the swelling of a nucleus to provide an environment in which all this synthetic activity can be initiated and sustained?

In conclusion, it seems appropriate to draw attention to the most recent explanation of the lampbrush phenomenon, offered by Anderson et al. (1982). The crucial and indisputable facts behind the Anderson hypothesis are these. First, the active transcription units of lampbrush chromosomes include genes for many species of maternal poly(A)mRNA. Second, there is no significant difference in structure between the poly(A)mRNA transcripts that have already accumulated by mid-oogenesis and those present in oocyte cytoplasm at ovulation. Third, newly synthesized (lampbrush) RNA contains repetitive sequence elements just as does the cytoplasmic RNA of late oocytes, and the sequence organization is the same in both cases. Fourth, the cytoplasmic poly(A)RNA is unstable, and continuous replacement is therefore needed throughout oogenesis. Finally, the turnover generated by loss and replacement of unstable message is a means of maintaining a large and diverse population of vital oocyte components over the long period of time when eggs are held in the ovary pending the appropriate conditions for ovulation.

Anderson et al. use these facts to support two conclusions. First, the observed rate and diversity of poly(A)RNA production in early stage oocytes demands maximum activity from all the transcription units on lampbrush chromosomes. Second, since these RNAs are abundant and unstable, an extraordinarily high rate of replacement synthesis is constantly needed throughout oogenesis. In essence, these authors are saying that lampbrush chromosomes generate the need for their own intense activity because they make so much and so many kinds of RNA in the first place, and they have to keep renewing this material for a long time afterwards. The idea is interesting and it should be tested, but only by persons who are keenly aware of the biochemistry *and* the biology of the organisms they employ.

REFERENCES

Anderson, D. M.; Richter, J. D.; Chamberlin, M. E.; Price, D. H.; Britten, R. J.; Smith, L. D.; and Davidson, E. H. (1982). Sequence organization of the poly (A) RNA synthesized and accumulated in lampbrush chromosome stage Xenopus laevis oocytes. *J. Mol. Biol.* **155**, 281-309.

Andreuccetti, P.; Taddei, C.; and Filosa, S. (1978). Intercellular bridges between follicle cells and oocyte during the differentiation of follicular epithelium in Lacerta sicula Raf. *J. Cell Sci.* **33**, 341-350.

Angelier, N.; and Lacroix, J. C. (1975). Complexes de transcription d'origines nucleolaire et chromosomique d'ovocytes de Pleurodeles waltlii et P.poireti (Amphibiens, Urodeles). *Chromosoma* **51**, 323-335.

Bakken, A. H.; and Graves, B. (1975). Visualisation of the tertiary structure of lampbrush chromosomes with the scanning electron microscope. *J. Cell. Biol.* **67**, 17a.

Baldwin, L.; and Macgregor, H. C. (1985). Centromeric satellite DNA in the newt Triturus cristatus karelinii and related species: Its distribution and transcription on lampbrush chromosomes. *Chromosoma* **92**, 100-107.

Barsacchi, G.; and Gall, J. G. (1972). Chromosomal localization of repetitive DNA in the newt Triturus. *J. Cell Biol.* **54**, 580-591.

Bona, M.; Scheer, U.; Bautz, E. F. K. (1981). Antibodies to RNA polymerase II(B) inhibit transcription in lampbrush chromosomes after microinjection into living amphibian occytes. *J. Mol. Biol.* **151**, 81-99.

Callan, H. G. (1957). The lampbrush chromosome of Sepia officinalis L, Anilocra physodes L and Scyllium catulus Cuv. and their structural relationship to the lampbrush chromosomes of Amphibia. *Pubbl. Staz. zool. Napoli* **29,** 329-346.
Callan, H. G. (1966). Chromosomes and nucleoli of the axolotl. Ambystoma mexicanum. *J. Cell Sci.* **1,** 85-108.
Callan, H. G. (1967). The organization of genetic units in chromosomes. *J. Cell Sci.* **2,** 1-7.
Callan, H. G. (1982). Lampbrush chromosomes. *Proc. R. Soc. Lond.* **B214,** 417-448.
Callan, H. G.; and Lloyd, L. (1960). Lampbrush chromosomes of crested newts Triturus cristatus (Laurenti). *Philos. Trans. R. Soc.* **B243,** 135-219.
Callan, H. G.; and Lloyd, L. (1975). Working maps of the lampbrush chromosomes of Amphibia. In *Handbook of Genetics* (R. C. King, ed.), vol. 4, pp. 57-77. Plenum Press, New York.
Callan, H. G.; and Macgregor, H. C. (1958). Action of deoxyribonuclease on lampbrush chromosomes. *Nature* (London) **181,** 1479-1480.
Davidson, E. H. (1976). *Gene activity in early development*. Academic Press, New York.
Diaz, M. O.; Barsacchi-Pilone, G.; Mahon, K. A.; and Gall, J. G. (1981). Transcripts of both strands of a satellite DNA occur on lampbrush chromosome loops of the newt Notophthalmus. *Cell* **24,** 649-659.
Duryee, W. R. (1937). Isolation of nuclei and non mitotic chromosome pairs from frog eggs. *Arch. exp. Zellforsch.* **19,** 171-176.
Duryee, W. R. (1941). The chromosomes of the amphibian nucleus. In *University of Pennsylvania Bicentennial Conference of Cytology, Genetics and Evolution*, pp. 129-141, University of Pennsylvania Press, Philadelphia.
Duryee, W. R. (1950). Chromosomal physiology in relation to nuclear structure. *Ann. N.Y. Acad. Sci.* **50,** 920-953.
Espinosa, A. M.; Leon, P. E.; Macaya, G.; Fuentes, A. L.; Gutierrez, J. M. (1980). Estructura y funcion de los cromosomas plumulados. *Rev. Biol. Trop.* **28,** 209-226.
Flemming, W. (1882). *Zellsubstanz, Kern und Zelltheilung*. F. C. W. Vogel, Leipzig.
Ford, P. J. (1971). Non-coordinated accumulation and synthesis of 5S ribonucleic acid by ovaries of Xenopus laevis. *Nature* (London) **233,** 561-563.
Gall, J. G. (1954). Lampbrush chromosomes from oocyte nuclei of the newt. *J. Morph.* **94,** 283-352.
Gall, J. G. (1955). Problems of structure and function in the amphibian oocyte nucleus. *Symp. Soc. Exp. Biol.* **9,** 358-370.
Gall, J. G. (1956). On the submicroscopic structure of chromosomes. *Brookhaven Symp. Biol.* **8,** 17-32.
Gall, J. G. (1963). Kinetics of deoxyribonuclease action on chromosomes. *Nature* (London) **198,** 36-38.
Gall, J. G.; Stephenson, E. C.; Erba, H. P.; Diaz, M. O.; Barsacchi-Pilone, G. (1981). Histone genes are located at the sphere loci of newt lampbrush chromosomes. *Chromosome* **84,** 159-171.
Gall, J. G.; Diaz, M. O.; Stephenson, E. C.; and Mahon, K. A. (1983). The transcription unit of lampbrush chromosomes. In *Gene Structure and Regulation in Development: Proceedings of the 41st Symposium of The Society for Developmental Biology*, pp. 137-146, Alan R. Liss, New York.
Goodpasture, C.; and Bloom, S. E. (1975). Visualization of nucleolar organizer regions in mammalian chromosomes using silver staining. *Chromosoma* **53,** 37-50.
Gould, D. C.; Callan, H. G.; and Thomas, C. A. (1976). The actions of restriction endonucleases on lampbrush chromosomes. *J. Cell Sci.* **21,** 303-313.
Harrison, C. J.; Allen, T. D.; and Harris, R. (1983). Scanning electron microscopy of variations in human metaphase chromosome structure revealed by Giemsa banding. *Cytogenet. Cell Genet.* **35,** 21-27.

Hartley, S. E.; and Callan, H. G. (1978). RNA transcription on the giant lateral loops of the lampbrush chromosomes of the American newt Notophthalmus viridescens. *J. Cell Sci.* **34,** 279-288.

Hilder, V. A.; Dawson, G. A.; and Vlad, M. T. (1983). Ribosomal 5S genes in relation to C-value in amphibians. *Nucl. Acids Res.* **11,** 2381-2390.

Hill, R. S. (1979). A quantitative electron-microscope analysis of chromatin from Xenopus laevis lampbrush chromosomes. *J. Cell Sci.* **40,** 145-169.

Hill, R. S.; and Macgregor, H. C. (1980). The development of lampbrush chromosome-type transcription in the early diplotene oocytes of Xenopus laevis: an electron-microscope analysis. *J. Cell Sci.* **44,** 87-101.

Jamrich, M.; Warrior, R.; Steele, R.; and Gall, J. G. (1983). Transcription of repetitive sequences on Xenopus lampbrush chromosomes. *Proc. Natl. Acad. Sci. (U.S.A.)* **80,** 3364-3367.

Kezer, J.; and Macgregor, H. C. (1973). The nucleolar organizer of Plethodon cinereus cinereus (Green). II. The lampbrush nucleolar organizer. *Chromosoma* **42,** 427-444.

Kleinschmidt, J. A.; Scheer, U.; Dabauvalle, M.-C.; Bustin, M.; Franke, W. W. (1983). High mobility group proteins of amphibian oocytes: a large storage pool of a soluble high mobility group-1-like protein and involvement in transcriptional events. *J. Cell. Biol.* **97,** 838-848.

Kloppstech, K. (1982). Acetabularia. In *The Molecular Biology of Plant Development* (H. Smith and D. Grierson, eds.), pp. 136-158. Blackwell, London.

Korn, L. J. (1982). Transcription of Xenopus 5S ribosomal RNA genes. *Nature* (London) **295,** 101-104.

Macgregor, H. C. (1972). The nucleolus and its genes in amphibian oogenesis. *Biol. Rev.* **47,** 177-210.

Macgregor, H. C. (1977). Lampbrush chromosomes. In *Chromatin and Chromosome Structure* (R. A. Eckhardt and Hsueh-Lei Li, eds.), pp. 339-357, Academic Press, New York.

Macgregor, H. C. (1980). Recent developments in the study of lampbrush chromosomes. *Heredity* **44,** 3-35.

Macgregor, H. C. (1982). Ways of amplifying ribosomal genes. In *The Nucleolus* (E. G. Jordan and C. A. Cullis, eds.), pp. 129-151. Cambridge University Press, Cambridge, England.

Macgregor, H. C. and Callan, H. G. (1962). The actions of enzymes on lampbrush chromosomes. *Quart. J. Microsc. Sci.* **103,** 173-203.

Macgregor, H. C.; and Kezer, J. (1970). Gene amplification in oocytes with 8 germinal vesicles from the tailed frog Ascaphus truei (Steineger). *Chromosoma* **29,** 189-206.

Macgregor, H. C.; and Klosterman, L. (1979). Observations on the cytology of Bipes (Amphisbania) with special reference to its lampbrush chromosomes. *Chromosoma* **72,** 67-87.

Macgregor, H. C.; and del Pino, E. M. (1982). Ribosomal gene amplification in multinucleate oocytes of the egg brooding hylid frog Flectonotus pygmaeus. *Chromosoma* **85,** 475-488.

Macgregor, H. C.; and Varley, J. M. (1983). *Working with Animal Chromosomes.* Wiley, New York.

Macgregor, H. C. and Uzzell, T. M. (1964). Gynogenesis in salamanders related to Ambystoma jeffersonianum. *Science* **143,** 1043-1045.

Mancino, G.; Nardi, I.; Corvaja, N.; Fiume, L.; and Marinozzi, V. (1971). Effects of α-amanitin on Triturus lampbrush chromosomes. *Exp. Cell Res.* **64,** 237-242.

Miller, D. A.; Dev, V. G.; Tantravahi, R.; and Miller, O. J. (1976). Suppression of human nucleolus organizer activity in mouse-human somatic hybrid cells. *Exp. Cell Res.* **101,** 235-245.

Miller, O. J.; Miller, D. A.; Dev, V. G.; Tantravahi, R.; and Croce, C. M. (1976). Expression of human and suppression of mouse nucleolus organizer activity in mouse-human somatic cell hybrids. *Proc. Nat. Acad. Sci. (U.S.A.)* **73,** 4531-4535.

Morgan, G. T.; Macgregor, H. C.; and Colman, A. (1980). Multiple ribosomal gene sites revealed by in situ hybridization of Xenopus rDNA to Triturus lampbrush chromosomes. *Chromosoma* **80**, 309-330.

Mott, M. R.; and Callan, H. G. (1975). An electron microscope study of the lampbrush chromosomes of the newt Triturus cristatus. *J. Cell Sci.* **17**, 141-261.

Old, R. W.; Callan, H. G.; Gross, K. W. (1977). Localization of histone gene transcripts in newt lampbrush chromosomes by in situ hybridization. *J. Cell Sci.* **27**, 57-79.

Olmo, E. (1983). Nucleotype and cell size in vertebrates: a review. *Bas. Appl. Histochem.* **27**, 227-256.

Perlman, S.; and Roshbash, M. (1978). Analysis of Xenopus laevis ovary and somatic cell polyadenylated RNA by molecular hybridization. *Develop. Biol.* **63**, 197-212.

Pukkila, P. J. (1975). Identification of lampbrush chromosome loops which transcribe 5S ribosomal RNA in Notophthalmus (Triturus) viridescens. *Chromosoma* **53**, 71-89.

Roeder, R. G. (1976). Eukaryotic nuclear RNA polymerases. In *RNA Polymerase* (R. Losick and M. Chaberlin, eds.), pp. 285-329, Cold Spring Harbor Laboratory, Cold Spring Harbor, New York.

Scheer, U. (1982). A novel type of chromatin organization in lampbrush chromosomes of Pleurodeles waltlii: visualization of clusters of tandemly repeated, very short transcriptional units. *Biol. Cell* **44**, 213-220.

Scheer, U.; Franke, W. W.; Trendelenburg, M. F.; and Spring, H. (1976). Classification of loops of lampbrush chromosomes according to the arrangement of transcriptional complexes. *J. Cell. Sci.* **22**, 503-519.

Scheer, U.; Spring, H.; and Trendelenburg, M. F. (1979a). Organization of transcriptionally active chromatin in lampbrush loops. In *The Cell Nucleus* (H. Busch, ed.) vol. 7, pp. 3-47, Academic Press, New York.

Scheer, U.; Sommerville, J.; and Bustin, M. (1979b). Injected histone antibodies interfere with transcription of lampbrush chromosome loops in oocytes of Pleurodeles. *J. Cell Sci.* **40**, 1-20.

Scheer, U.; Hinssen, H.; Franke, W. W.; and Jockusch, B. M. (1984). Involvement of nuclear actin in transcription of lampbrush chromosomes of amphibian oocytes demonstrated by microinjection of actin-binding proteins and actin antibodies. *Cell* **39**, 111-122.

Shiokawa, K. (1983). Mobilization of maternal mRNA in amphibian eggs with special reference to the possible role of membranous supramolecular structures. *FEBS. Letters* **151**, 179-184.

Sims, S. H.; Macgregor, H. C.; Pellatt, P. A.; and Horner, H. A. (1984). Chromosome 1 in crested and marbled newts (Triturus). An extraordinary case of heteromorphism and independent chromosome evolution. *Chromosoma* **89**, 169-185.

Sommerville, J. (1977). Gene activity in the lampbrush chromosomes of amphibian oocytes. *Int. Rev. Biochem.* **15**, 79-156.

Sommerville, J. (1981). Immunolocalization and structural organization of nascent RNP. In *The Cell Nucleus* (H. Busch, ed.), vol. VIII, pp. 1-57. Academic Press, New York.

Spring, H.; Scheer, U.; Franke, W. W.; and Trendelenburg, M. F. (1975). Lampbrush-type chromosomes in the primary nucleus of the green alga Acetabularia mediterranea. *Chromosoma* **50**, 25-43.

Thomas, T. L.; Posakony, J. W.; Anderson, D. H.; Britten, R. J.; and Davidson, E. H. (1981). Molecular structure of maternal RNA. *Chromosoma* **84**, 319-335.

Tomlin, S. G.; and Callan, H. G. (1951). Preliminary account of an electron microscope study of chromosomes from newt oocytes. *Quart. J. Microsc. Sci.* **92**, 221-224.

Trendelenburg, M. F.; Spring, H.; Scheer, U.; and Franke, W. W. (1974). Morphology of nucleolar cistrons in a plant cell, Acetabularia mediterranea. *Proc. Nat. Acad. Sci. (U.S.A.)* **71**, 3626-3630.

Varley, J. M..; and Morgan, O. T. (1978). Silver staining of the lampbrush chromosomes of Triturus cristatus carnifex. *Chromosoma* **67**, 233-244.

Varley, J. M.; Macgregor, H. C.; Nardi, I.; Andrews, C.; and Erba, H. P. (1980). Cytological evidence of transcription of highly repeated DNA sequences during the lampbrush stage in Triturus cristatus carnifex. *Chromosoma* **80**, 289-307.

Vlad, M. T.; and Macgregor, H. C. (1975). Chromomere number and its genetic significance in lampbrush chromosomes. *Chromosoma* **50**, 327-347.

6

Dosage Compensation in Mammals: X Chromosome Inactivation

Leslie F. Lock and Gail R. Martin
University of California at San Francisco
San Francisco, California

Evolution of heteromorphic sex chromosomes has resulted in disparity between males and females with respect to the number of copies of X-linked genes in a diploid complement. In species with heteromorphic sex chromosomes, the homogametic sex (XX or ZZ) has two copies of X (Z) linked genes, whereas the heterogametic sex (XY or ZW) has only one. In spite of this inequity, however, in many of these species the level of expression of X-linked genes is equal in the adult female and male. This paradox was first described by Muller (1950) following his studies of an X-linked eye pigment gene, white, in *Drosophila*. He observed that females homozygous for a defective allele are identical in eye color to males hemizygous for the defective allele. This suggested that an equivalent amount of pigment was produced by two copies of the mutant gene in females as by one copy of the mutant gene in males. Muller proposed that X-linked genes are "dosage compensated" such that two X chromosomes in females and one X chromosome in males produce equivalent levels of X-linked gene product.

In the 35 years since the concept of dosage compensation was first suggested, numerous studies of this phenomenon have been carried out on a great variety of organisms. It is now clear that in several species of birds, reptiles and butterflies, there is no dosage compensation of Z-linked genes (Ohno, 1967; Johnson and Turner, 1979; Baverstock et al., 1982). Thus, the homogametic sex (ZZ) expresses twice the level of Z-linked gene products as does the heterogametic sex (ZW). Apparently, Z-linked genes in these species are either (1) functionally dosage independent, so that a two fold difference in the level of these gene products has no effect or (2) sex-specific in function. In contrast to the absence of dosage compensation in some species, some form of dosage compensation has been observed in such diverse organisms as nematodes, insects and mammals (Muller, 1950; Lyon, 1961; Meneely and Wood, 1984). The variety of mechanisms that have evolved reflects the diversity of species that undergo dosage compensation. The differences in the strategies of dosage compensation observed in *Drosophila* and mammals illustrate this point. In *Drosophila*, equivalent levels of X linked gene products in male (XY) and female (XX) result from unequal transcription of X-linked genes in the two sexes (Kazazian et al., 1965; Seecof et al., 1969). Although both X chromosomes in females are transcribed, transcription from the single X in males is equivalent to that from the two X chromosomes in females (Mukherjee and Beermann, 1965). In contrast, in mammals one of the two X chromosomes in each female cell is inactive, not expressed at all. Equivalent levels of X-linked gene products result from transcription of a single X in both males and females. Thus, dosage compensation in these two species is accomplished by quite different mechanisms: in one, transcription is altered so that equivalent levels of product result from unequal amounts of genetic material, whereas in the other, the net amount of genetic material available for transcription is altered.

The latter form of dosage compensation, known as X chromosome inactivation, is the subject of this review. The evidence in support of X inactivation in mammals, as hypothesized by Lyon (1961; 1962) and Beutler et al. (1962), observations that have led to modifications and extensions of the original hypothesis, and possible mechanisms of X inactivation will be discussed.

SINGLE ACTIVE X HYPOTHESIS

In 1961, M. F. Lyon proposed that one of the two X chromosomes in each cell of female mammals is inactive. This "single active X" hypothesis was based on several observations. First, XO female mice are virtually normal, indicating that the second X of female mice is expendable. Second, the sex chromatin body found in the nuclei of certain cell types of the female mammal had been shown to consist of a single heterochromatic X chromosome (Ohno, et al., 1959). Third, female mice heterozygous for certain X linked coat color mutations have a variegated coat pattern consisting of phenotypically normal and mutant patches. Lyon suggested that patches with a mutant phenotype are derived from clones of cells in which the X chromosome carrying the normal allele was inactivated. Conversely, the phenotypically normal patches represent clones in which the X chromosome carrying the mutant allele was inactivated. Thus, the female mammal would be a mosaic, consisting of two populations of cells, each with a different X chromosome inactivated.

The single active X hypothesis as proposed by Lyon (1961, 1962) consists of four major tenets:

1. one of the two X chromosomes of female mammals is inactive;
2. inactivation takes place early in embryonic development;
3. the inactivated X can be either paternal or maternal in origin, with the choice being made at random;
4. after inactivation of one of the X chromosomes occurs in a cell, the inactive X is stable and inherited by all descendants of that cell.

Since the proposal of this hypothesis, considerable evidence has been amassed in support of these four major tenets, and only a few modifications of the original proposal have been necessary.

One Active and One Inactive X Chromosome Is Present in Each Cell of Normal Female Mammals

The first major tenet of Lyon's hypothesis is that only one of the two X chromosomes in each cell of female mammals is expressed; the other X is inactive. Both cytological and biochemical evidence support this point. First, there are several cytological differences between the two X chromosomes in females that are consistent with a difference in transcriptional activity. One of the X chromosomes is heterochromatic, replicates late in the synthetic phase of the cell cycle, and incorporates little ^{3}H-uridine (Ohno and Hauschka, 1960; Morishima et al., 1962; Mukherjee and Sinha, 1964; Evans et al., 1965; Comings, 1966a). In contrast, the other X is euchromatic, replicates in concert with the autosomes, and incorporates ^{3}H-uridine at a level similar to that of the autosomes, indicating a normal level of transcription. Second, female humans appear to be mosaics in terms of the expression of X-linked genes. Females heterozygous for X-linked glucose-6-phosphate dehydrogenase (G6PD) deficiency have two populations of red blood cells, half completely normal and half completely deficient for G6PD (Beutler et al., 1962). Similar analysis of hypoxanthine phosphoribosyltransferase (HPRT) activity in clones of skin fibroblasts from human females heterozygous for HPRT deficiency also indicates the existence of two populations of cells, one with a normal level of HPRT activity and one with no HPRT activity (Migeon et al., 1968). Further evidence that only one X is active in each female cell is provided by studies of the expression, in single cell clones isolated from female mammals, of X-linked genes encoding electrophoretic variants of certain enzymes. Single cell clones of cultured cells derived from a female human heterozygous for an electrophoretic variant of G6pd expressed either one or the other allele, but never both (Davidson et al., 1963). Similar studies of female humans heterozygous for variants of two X-linked loci, (G6pd and phosphoglycerate kinase-1, Pgk-1 or G6pd and Hprt), have shown expression of alleles in *cis* indicating that an entire X-chromosome is inactivated; see Figure 6-1 (Chen et al., 1971; Goldstein et al., 1971; Gartler et al., 1972a).

In contrast to the relative abundance of variants of X-linked enzymes in humans, variants of X-linked enzymes in other mammalian species are fairly uncommon. Consequently, most studies aimed at demonstrating mosaicism of X-linked gene expression at the cellular level in mammalian species other than humans have been performed using interspecific hybrids, such as hybrids between related hare species or hybrids between related mouse species. These have confirmed that gene products coded for by only one of the X chromosomes are expressed in individual cells of females (Mukherjee

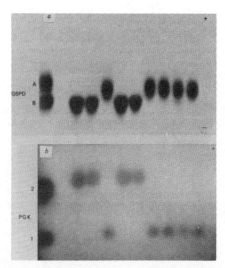

Figure 6-1. Expression of G6PD and PGK-1 in single cell clones derived from a human female heterozygous for both X-linked genes.
Expression of alleles in *cis* (PGK 1^1, G6PD A or Pgk 1^2, G6PD B) in all clones examined suggests that there is extensive inactivation of one of the two X chromosomes in female humans. The left lane shows the heterozygous control; the remaining lanes show different clones. *(From Gartler, et al., 1972a, p. 149; copyright © 1972 by Macmillan Journals Ltd.; reprinted by permission.)*
a) G6PD types.
b) PGK-1 types.

and Sinha, 1964; Ohno et al., 1965; Cohen and Ratazzi, 1971; Ray et al., 1972; Chapman and Shows, 1976). These observations indicate that only one of the two X chromosomes in each cell of female mammals is active and that the other is inactive, supporting the first major tenet of the single X active hypothesis.

An apparent exception to the rule of inactivation of one of the two X chromosomes in females has been observed in studies of the expression of two X-linked loci, the blood antigen, Xg, and steroid sulfatase (Sts), in single cell clones from human females (Fialkow, 1970; Fialkow et al., 1970; Shapiro et al., 1979; Mohandas et al., 1980). All single cell clones from female humans heterozygous for deficient alleles of either Xg or Sts express the loci. Also, analysis of clones from double heterozygotes for HPRT deficiency and STS deficiency indicate that two populations do exist; one deficient for HPRT and one with HPRT activity. Both populations, however, express STS (Shapiro et al., 1979). These data indicate that STS activity is expressed by the normal STS allele irrespective of whether it is located on the active or inactive X chromosome. Additional evidence for expression of the STS gene on the inactive X comes from analysis of a mouse/human somatic cell hybrid line that retains an inactive X chromosome but no active X chromosome. This cell line expresses human STS, presumably from the STS gene on the inactive human X (Mohandas et al., 1980). Since the Xg and STS genes are located within 10 centimorgans of one another on the distal half of the short arm of the human X chromosome, it has been suggested that an entire region of the short arm of the human X escapes inactivation (Race and Sanger, 1975; Mohandas et al., 1979; Muller et al., 1981).

Although it has not been definitively proven, it now appears that the STS locus also escapes inactivation in other mammalian species. Data consistent with those from studies of humans have been obtained from studies of STS expression in wood lemmings (Ropers and Wiberg, 1982). In addition, despite earlier studies suggesting that STS levels are similar in XX and XO female mice and XY male mice (Crocker and Craig, 1983; Gartler and Rivest, 1983), more recent studies indicate that the STS gene does not undergo inactivation in mice, furthermore, it has been suggested that in mice a functional STS allele exists on the Y chromosome and that it undergoes obligatory recombination during meiosis with the X-linked allele (Keitges et al., 1985). This hypothesis was proposed to explain data indicating that STS is X-linked despite its apparently autosomal pattern of inheritance in mice, and the finding that the levels of STS activity are the same in tissues from females heterozygous for an STS-deficient allele ($X+/X-$), XO females carrying a wild-type allele of STS ($X+/O$) and males derived from STS-deficient fathers ($X+/y-$).

Although expression of the STS allele on the inactive X in several species seems likely, there are observations that indicate that there are pecularities about this apparent exception to the one active X rule. The level of STS expression from the inactive X does not appear to be as high as that from the active X. In human cells, the ratio of STS activity between females and males ranges from 1.0 to 1.7 (Muller et al., 1981; Ropers et al., 1981; Shapiro et al., 1982; Chance and Gartler, 1983). Also, cells with supernumerary X chromosomes (up to 4X) have only twice the level of STS as males (XY), instead of a level reflective of the number of X chromosomes (STS genes) present (Ropers et al., 1981; Chance and Gartler, 1983). In addition, analysis of STS activity in single cell clones derived from females heterozygous at two X linked loci, G6pd and STS, indicates that STS activity from the normal allele is significantly lower when it is on the inactive as compared with on the active X (Migeon et al., 1982). These studies suggest that in human cells either (1) STS is partially inactivated on the inactive X or (2) the STS gene on the active X is up regulated; in either case the net effect is a higher level of expression by the STS gene on the active X as compared to that on the inactive X.

X Inactivation Occurs During Early Embryogenesis

The second tenet of the single active X hypothesis is that the process by which one of the two X chromosomes in each cell of female mammals becomes inactive takes place early in embryonic development. Attempts to demonstrate this have been concerned not only with the question of when it occurs but also with the question of whether the presence of one active and one inactive X in females results from the inactivation of one of two previously active X chromosomes or whether it results from the activation of one of two previously inactive X chromosomes.

Most of the knowledge about X inactivation comes from studies of the mouse embryo. An understanding of the experiments that have been performed requires some knowledge of the sequence of events and the lineage relationships of the cells of the early mouse embryo (Gardner and Papaioannou, 1975). These are illustrated in Figure 6-2. Briefly, at approximately 24 hours after fertilization, cleavage begins in the mouse embryo. When the embryo consists of 8 to 16 cells, a process known as compaction occurs and the morula is formed. Subsequently, at about 3.0 days after fertilization, the outer cells of the compacted morula differentiate to a cell type known as trophectoderm and a fluid-filled cavity, the blastocoel, forms in the interior of the embryo. The remaining inner cells form the inner cell mass (ICM). The trophectoderm is an extraembryonic cell type, whereas the ICM consists of the pluripotent cells that will eventually form the fetus. At about 4.0 days of development, the primitive endoderm forms from the cells on the blastocoelic surface of the ICM. Like the trophectoderm, these differentiated endodermal cells also contribute exclusively to extraembryonic tissues. The interior cells of the ICM, now refered to as primitive ectoderm, remain pluripotent. At about 5.0 to 5.5 days of development, however, gastrulation begins and the cells of the primitive ectoderm begin their differentiation into the three germ layers that form the fetus: mesoderm, definitive ectoderm and definitive endoderm.

As yet, significant activity of the two X chromosomes in female embryos has not been demonstrated during the early cleavage stages. Instead the level of X-linked enzymes in embryos at the 1 to 8 cell stage reflects the chromosome constitution of the mother. This indicates that most, if not all, X-linked enzyme activity in the earliest stages of development is maternal, not embryonic, in origin (Epstein, 1972; Kozak and Quinn, 1975; Monk and Kathuria, 1977; Kratzer and Gartler, 1978; Monk and Harper, 1978). In contrast,

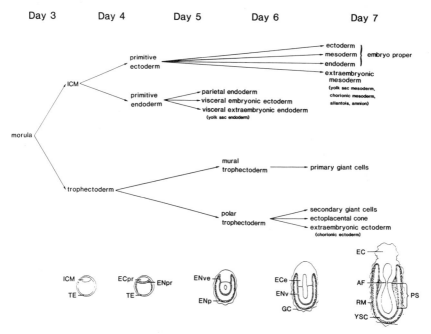

Figure 6-2. Embryonic stages, time post-fertilization, and cell lineages of the mouse embryo.
ICM, inner cell mass; TE, trophectoderm; ECpr, primitive ectoderm; ENpr, primitive endoderm; ENve, visceral extraembryonic endoderm; ENp, parietal endoderm; ECe, extraembryonic ectoderm; ENv, visceral endoderm; GC, giant cells; EC, ectoplacental cone, AF, amniotic fold; RM, Reichert's membrane; YSC, yolk sac cavity; PS, primitive streak. *(After Beddington, 1983, p. 3; copyright © 1983 by Elsevier Biomedical Press B.V.)*

several observations suggest that both X chromosomes are active in morula stage embryos. First, in morula stage embryos both X chromosomes are euchromatic and replicate in synchrony with the autosomes (Kinsey, 1967; Issa et al., 1969; Takagi, 1974). This is consistent with both X chromosomes being active. Second, female embryos at the morula stage have twice the level of HPRT as do males (Epstein, 1972; Kratzer and Gartler, 1978; Monk and Harper, 1978). Taken together, these observations suggest that by the morula stage both X chromosomes are active in embryonic cells, therefore the process that results in only one active X chromosome at subsequent stages must be one of inactivation rather than activation of one of two previously inactive X chromosomes.

Further evidence that there is, in fact, an X inactivation process comes from the studies of embryonal carcinoma (EC) cell lines. EC cells are the stem cells of teratocarcinomas, which are tumors that either arise spontaneously in the gonads of certain strains of mice or are induced experimentally by transplantation of an early embryo to an extrauterine site. There is considerable evidence that these cells are analogous to the cells of normal embryonic ICM or primitive ectoderm (Martin, 1980). Studies of both gene dosage of X-linked enzymes and replication pattern of the X chromosomes have shown that both X chromosomes can be active in undifferentiated female EC cells and that during differentiation *in vitro* these cells can undergo X inactivation as demonstrated by occurrence of dosage compensation and/or the appearance of an asynchronously replicating X chromosome (Martin et al., 1978; McBurney and Strutt, 1980; Takagi and Martin, 1984).

Given that inactivation of one of two active X chromosomes occurs during early embryogenesis, the question remains whether this process occurs in all cells of the embryo simultaneously. It has now been demonstrated that in fact X inactivation takes place in three to four waves, concomitant with the differentiation of various cell lineages in the embryo (Monk, 1978; Monk and Harper, 1979). Evidence that X inactivation has occurred is first demonstrable in the blastocyst by the presence of a heterochromatic, asynchronously labeling X (Issa et al., 1969; Takagi, 1974) and dosage compensation of X-linked enzymes (Monk and Kathuria, 1977; Epstein et al., 1978; Kratzer and Gartler, 1978; Monk and Harper, 1979). These observations, however, appear to reflect the occurrence of X inactivation in the trophectoderm only; cells of the ICM do not undergo X inactivation at the blastocyst stage. Activity of both X chromosomes in the ICM was first demonstrated by Gardner and Lyon (1971). Chimeric mice were produced by injection of a single ICM cell from a day 3.5 female embryo heterozygous for an X-linked coat color mutation into a recipient blastocyst carrying a third allele at the coat color locus. Some of the resultant chimeras were tricolored with patches reflecting expression of the allele carried by cells of the recipient blastocyst, as well as both alleles carried on the X chromosomes of the single injected ICM cell. This suggests that at the time they were injected into the recipient blastocyst at least some of the ICM cells in the donor blastocyst had two active X chromosomes. The idea that ICM cells do have two active X chromosomes has been confirmed both biochemically (ICM cells are not dosage compensated, Monk and Harper, 1979), and cytogenetically (ICM cells do not have an asynchronously labeling, inactive X chromosome, Takagi et al., 1982). Thus, at the blastocyst stage, the embryo consists of two populations of cells: the ICM, a pluripotent cell population which has not undergone inactivation, and the trophectoderm, a differentiated cell population which has undergone inactivation.

The second instance of inactivation occurs at about 5.0 days of development, concomitant with differentiation of primitive endoderm. At this stage both X chromosomes remain active in the pluripotent cell population, the primitive ectoderm (Monk and Harper, 1979; Takagi et al., 1982). Subsequently, at about six days of development, this pluripotent population is the third to undergo inactivation as determined by studies of dosage compensation (Monk and Harper, 1979) as well as the appearance of the allocyclic X (Takagi et al., 1982). From these studies, however, it has not been possible to ascertain whether all cells of the primitive ectoderm undergo inactivation at this time. The results of several studies of the distribution and patch size of melanocytes in females chimeric or heterozygous for X-linked genes appear to be consistent with inactivation occurring in subpopulations of the primitive ectoderm from day 6.0 to 13.0 (Deol and Whitten, 1972*a*, 1972*b*; McLaren et al., 1973). In one such study, the distribution of pigmented and unpigmented cells in the retina was compared in (1) females heterozygous at an X-linked pigment gene, and (2) chimeric females produced by aggregation of two embryos homozygous for different alleles at the X-linked pigment locus. Since the size of pigmented patches in heterozygous females was much smaller than that in chimeric females, it was concluded that X inactivation in the female heterozygotes must have occurred at an embryonic stage later than the stage at which the chimeras were formed (morula stage). Furthermore, from the extent of the difference in patch size, it was estimated that inactivation occurred in the precursors of pigmented retinal cells at day 6.0. Estimates from similar analyses have placed the time of inactivation at 13.0 and 6.0 days post-fertilization for melanocytes of the inner ear and tail skin, respectively. However, such estimates based on patch size in heterozygous females and chimeric females presuppose that there are no significant differences in cell mixing during embryogenesis in heterozygous female embryos as compared to chimeric female embryos. This assumption may not be correct, and it is possible that the observed difference in patch size reflects decreased mixing of cells of different genotypes in the chimeric embryos (McLaren et al., 1973).

It has also been suggested that X inactivation may not occur in germ cell progenitors at gastrulation (day 6.0). Although the time at which X inactivation occurs in germ cell precursors is still unknown, the controversial question as to whether X inactivation occurs in germ cells has been resolved. In the 12 week human fetus, the absence of a heterodimer in germ cell progenitors from females heterozygous for electrophoretic variants of G6PD indicates that X inactivation has occurred in these germ cell progenitors (Gartler et al., 1975). In 12 day fetal mice, the presence of an asynchronously labeling X and dosage compensation of HPRT in germ cell progenitors indicate that X inactivation has occurred (Gartler et al., 1980; Johnston, 1981; McMahon et al., 1981; Monk and McLaren, 1981). Finally, in *Mus caroli*, a related species of mice, the heterodimer of G6PD is not present in germ cell progenitors isolated from females heterozygous for electrophoretic variants of G6pd (Kratzer and Chapman, 1981). These observations indicate that X inactivation has occurred in germ cell progenitors by day 12 in mice, prior to or during the mitotic phase which precedes entry into meiosis. Since it has been shown that both X chromosomes function in oocytes (post-entry into meiosis) in mice (Epstein, 1969, 1972; Kozak et al., 1974; Mangia et al., 1975; Monk and Kathuria, 1977) and humans (Gartler et al., 1972b, 1973, 1975), the inactive X present in mitotic germ cell precursors must be reactivated at some point. Detailed studies of premeiotic and meiotic germ cells in mouse and human at both the cytological and biochemical level have suggested that the two active X chromosomes are present in the germ cells during meiotic prophase and that reactivation of the inactive X occurs just prior to entry into meiosis (Mangia et al., 1975; Gartler et al., 1975; Johnston, 1981; Kratzer and Chapman, 1981; McMahon et al., 1981; Monk and McLaren, 1981).

X Inactivation Can Be Random or Nonrandom

Random Inactivation. The third tenet of the single active X hypothesis is that the X chromosome from either parent can be inactivated, and that the choice is made randomly. Numerous studies of the expression of X-linked genes in adult mammals have confirmed that the maternally and paternally derived X chromosomes are inactivated in equal proportions in adult tissues. For example, the maternally and paternally derived X chromosomes are expressed in equal proportions of single cell clones derived from human females heterozygous for variants of G6PD, HPRT, or PGK-1 (Davidson, 1963; Migeon, et al. 1968; Chen et al., 1971; Goldstein et al., 1971; Gartler et al., 1972a). In addition, in adult tissues from mice carrying two cytogenetically distinguishable X chromosomes (one normal X and one abnormally long X, Cattanach's translocation) the normal and abnormal X chromosomes are late labeling in equal proportions of cells (Evans et al., 1965). Thus, the third tenet of the single active X hypotheisis has been confirmed for normal adult tissues; either X chromosome can be inactivated, with the choice being made at random.

Nonrandom Inactivation. In contrast to the random inactivation of one of the two X chromosomes observed in tissues of normal adult humans and mice, several instances of nonrandom expression of one of the two X chromosomes have been observed. Such instances fall into roughly two categories: nonrandom expression of one of two X chromosomes that differ genetically, either at the single locus or gross chromosomal level, and nonrandom expression of one of two X chromosomes as a function of their parental origin.

Nonrandom Expression as a Function of Genetic Differences Between the Two X Chromosomes. Structural abnormalities of one of the two X chromosomes can lead to nonrandom expression of either the normal or abnormal X chromosome. In female humans carrying an X chromosome with a region deleted, the deleted X chromosome is always inactive, probably to avoid nullisomy for the deleted genes that would result from inactivation of the normal X (Gartler, 1976; Latt et al., 1976). In the case of most X/autosome translocations in humans, the normal X chromosome is usually inactive. It has been suggested that this may be a consequence of detrimental effects on cell survival resulting from the spread of inactivity from the X region to the autosomal portion in the X/autosome translocation (Gartler, 1976; Latt et al., 1976). Similarly, for at least one X/autosome translocation in the mouse there is nonrandom expression of the translocated X (Searle, 1962; Lyon et al., 1964; Takagi, 1980; Disteche, 1981). However, many X/autosome translocations in mice undergo random inactivation, some resulting in variegated coat color due to the spread of inactivity from the X portion of the chromosome to the autosomal loci near the breakpoint of the translocation (Russell, 1961, 1963; Eicher, 1970).

In addition to such gross structural abnormalities, genetic deficiencies at a single locus can also lead to nonrandom expression of one of the two X chromosomes, at least in some tissues. For example in female humans heterozygous for HPRT deficiency, there is evidence that, although either X chromosome, the one carrying the normal or the deficient Hprt gene, is expressed in fibroblasts, only the X chromosome with the normal Hprt gene is active in erythrocytes (Migeon et al., 1968; Salzman et al., 1968; Nyhan et al., 1970; Migeon, 1970). Nonrandom expression of the X chromosome carrying the normal allele is also observed in erthrocytes of some female humans heterozygous for G6PD deficiency (Gandini et al., 1968).

Variation at a single gene locus, X chromosome controlling element, Xce, can also cause nonrandom expression of one of the two X chromosomes (Cattanach and Isaacson, 1967). Several alleles at this locus have been identified in mice by their effects on the extent of variegation of coat color in females heterozygous for X-linked coat color mutations or on the extent of nonrandom expression of the X-linked gene Pgk-1 (Cattanach et al., 1969, 1970; Johnston and Cattanach, 1981; West and Chapman, 1978). It has been shown by analysis of the expression of X-linked genes in female offspring from reciprocal crosses that the nonrandom expression caused by variants of Xce is independent of parental origin (Cattanach and Isaacson, 1967; Johnston and Cattanach, 1981; West and Chapman, 1978; Rastan, 1982).

A fourth case of nonrandom expression of genetically different X chromosomes occurs in interspecific hybrids. Several examples have been observed. In crosses of horse and donkey, the fact that the horse and donkey X chromosomes differ cytogenetically and biochemically has made possible analysis of X chromosome expression in the interspecific hybrids that result. Analysis of G6PD expression and replication patterns of the X chromosomes indicate that the donkey X chromosome is preferentially inactive; see Figure 6-3 (Cohen and Ratazzi, 1971; Gianelli and Hamerton, 1971; Hamerton et al., 1971; Mukherjee and Milet, 1972; Ratazzi and Cohen, 1972; Ray et al., 1972). A similar observation has been made in interspecific fox hybrids (Serov et al., 1978). In contrast, in interspecific hybrids between two related mouse species, *Mus musculus* and *Mus caroli*, nonrandom expression of one of the two X chromosomes has not been observed (Chapman and Shows, 1976). Thus, genetic differences between two X chromosomes, including obvious deficiencies such as chromosomal and single gene abnormalities as well as the undetermined differences between X chromosomes from different species, can result in nonrandom expression of one of the two X chromosomes.

Nonrandom Expression as a Function of Differences in Parental Origin of the Two X Chromosomes. The second category of nonrandom expression of X chromosomes involves nonrandom expression of apparently identical X chromosomes on the basis of parental origin. This has been observed in two instances. First, marsupials normally express only the maternally derived X chromosome in most tissues (Cooper et al., 1971; Sharman, 1971;

Vandeberg et al., 1973; Johnston et al., 1978). This was determined both biochemically in studies involving electrophoretic variants of X-linked genes and cytogenetically in studies involving cytogenetically different X chromosomes. That this nonrandom expression of one of the two X chromosomes was based on parental origin rather than genetic differences between the two X chromosomes was determined by analysis of heterozygous females derived from reciprocal crosses (Johnston et al., 1978). The maternally derived X chromosome was preferentially expressed regardless of its genetic constitution.

Similar nonrandom expression of the maternally derived X chromosome is observed in several extraembryonic lineages in mice and rats. This was first demonstrated by Takagi and Sasaki (1975), using embryos with one normal and one abnormally long X chromosome (Cattanach's translocation). In these embryos the paternally derived X was asynchronously replicating in greater than 90% of the cells of the extraembryonic portion of the day 6.5 embryo. Extraembryonic tissues, yolk sac and chorion, of the day 7.5 embryo also showed preferential inactivation of the paternally derived X. Conversely, in fetal lineages of both the day 6.5 and day 7.5 embryo, the maternally and paternally derived X chromosomes were asynchronously labeled in equal proportions of cells. These observations have been confirmed and extended with biochemical studies in mouse embryos heterozygous for an electrophoretic variant of Pgk-1 (West et al., 1977; Frels et al., 1979; Frels and Chapman, 1980; Papaioannou et al., 1981; Harper et al., 1982) and with cytogenetic studies (Wake et al., 1976; Takagi, 1978; Takagi et al., 1978). That this

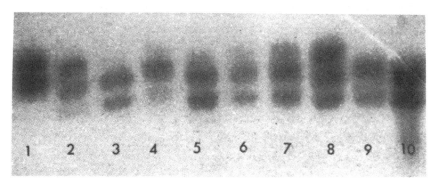

Figure 6-3. Expression of G6PD in female interspecific hybrids between horse and donkey.
The relative level of expression of the donkey and horse forms varies between individuals and among tissues of a single individual. Lanes 1 and 10 are donkey and horse standards, respectively; lanes 2 to 9 are pancreas and parotid gland samples from several individuals. *(Reproduced from Hook and Brustman, 1971, p. 349; copyright © 1971 by Macmillan Journals Ltd.; reprinted by permission.)*

nonrandom expression is a consequence of the parental origin of the X chromosome has been demonstrated by analysis of embryos heterozygous for electrophoretic variants of Pgk-1 derived from reciprocal crosses; see Figure 6-4 (West et al., 1977; Frels et al., 1979; Papaioannou and West, 1981; Harper et al., 1982). The maternally derived X is preferentially expressed regardless of the Pgk-1 variant it carries. Reciprocal crosses with cytogenetically marked X chromosomes (e.g., Cattanach's translocation) also indicate that nonrandom expression of one of the two X chromosomes in extraembryonic tissues of mice and rats is dependent on maternal origin rather than genetic constitution (Takagi and Sasaki, 1975; Wake et al., 1976). These studies indicate that the paternally derived X chromosome is preferentially inactivated in trophectoderm and its derivatives (ectoplacental cone and extraembryonic ectoderm: chorionic ectoderm) and primitive endoderm and its derivatives (parietal endoderm, visceral embryonic endoderm, and visceral extraembryonic endoderm: yolk sac endoderm). In contrast, the maternally and paternally derived X chromosomes are inactivated with equal frequency in embryonic ectoderm and its derivatives which form the fetus and contrib-

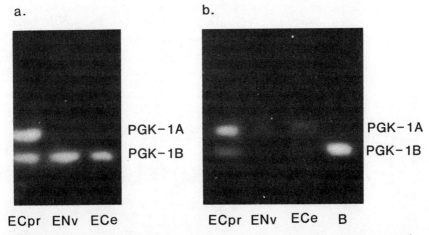

Figure 6-4. Expression of PGK-1 in tissues of a day 6.5 mouse embryo heterozygous for electrophoretic variants of PGK-1.
Expression of only the maternally derived form of PGK-1 is detectable in visceral endoderm and extraembryonic ectoderm, whereas the maternally and paternally derived forms both are expressed in primitive ectoderm. ECpr, primitive ectoderm; ENv, visceral endoderm; ECe, extraembryonic ectoderm; B, PGK-1B standard. *(Reprinted from Harper et al., 1982, copyright © 1982 by Company of Biologists Ltd.)*
a) Female embryo from the cross: Pgk-1^b/Pgk-1^b female × Pgk-1^a male.
b) Female embryo from the cross: Pgk-1^a/Pgk-1^a female × Pgk-1^b male.

ute to some extraembryonic tissues (extraembryonic mesoderm: yolk sac and chorionic mesoderm, amnion, allantois).

Causes of Nonrandom Expression of One of the Two X Chromosomes in Females: Random Inactivation Followed by Selection or Primary Nonrandom Inactivation?

Apparent nonrandom X inactivation can result from either initial random inactivation followed by selective growth of the cells with the given X active, or initial nonrandom inactivation. In general, the apparent nonrandom X inactivation associated with genetic differences between the two X chromosomes is thought to result from initial random inactivation followed by selective loss of cells with the "wrong" X inactive (Lyon et al., 1964; Gartler and Sparkes, 1963). Although this has not been definitively proven, several observations are consistent with this hypothesis. First, selection in vitro on the basis of genetic differences, including HPRT activity, has been demonstrated (Felix and DeMars, 1969; Salzman et al., 1968; Migeon, 1978). Also, selection appears to be occurring in vivo, as evidenced by the progressive increase in the extent of nonrandomness in X inactivation during embryogenesis and/or aging (Albertini and DeMars, 1974; Migeon, 1978; Disteche et al., 1979; Takagi, 1980). Selection during embryogeneis is thought to account for the small size of human fetuses carrying an isochromosome X (Gartler and Sparkes, 1963). Finally, the rarity of complete nonrandom expression and variations from tissue to tissue in the extent of nonrandom expression suggests that it results from selection and that the selection pressures differ in different tissues (Hook and Brustman, 1971; Latt et al., 1976; Russell and Cachiero, 1978). Such selective differences could result from many factors, including differing degrees of metabolic cooperation between cells in different tissues (Nyhan et al., 1970; Gartler, 1976; Migeon, 1978).

One exception to this generalization is the nonrandom expression caused by allelic variants of Xce. Nonrandom expression of one of the two X chromosomes appears to result from primary nonrandom inactivation, since nonrandom expression, equivalent in extent to that observed in adults, can be detected in the embryo as early as 6.5 days of development (West and Chapman, 1978; Johnston and Cattanach, 1981; Rastan, 1982).

The preferential expression of the maternal X chromosome is trophectoderm and primitive endoderm and their derivatives, at least in mice, also appears to result from initial nonrandom inactivation. This is evidenced by the fact that the maternally derived form of Pgk-1 predominates in extraembryonic tissues immediately after inactivation has occurred in those tissues (Takagi et al., 1978; Frels and Chapman, 1980; Papaioannou et al., 1981; Harper et al., 1982), and the time elapsed is insufficient to allow for selective growth. In addition, studies indicate that there is no selective growth of cells with the maternal X active as a consequence of the maternal environment. When Pgk-1 a/b heterozygous female embryos are transfered from their natural mother homozygous for Pgk-1a to foster mothers homozygous for Pgk-1b, the maternal form of PGK-1 derived from the natural mother (Pgk-1a) still predominates in extraembryonic tissues (Frels et al., 1979; Papaioannou and West, 1981).

X Chromosome Inactivity Is Stable and Heritable

The fourth major tenet of the single active X hypothesis as proposed by Lyon (1961), is that once X inactivation has occurred in a cell, the inactivity of the X is stable and inherited by all descendants of that cell. Numerous studies indicate that, with the exception of the germ cells, repression of the inactive X chromosome is relatively stable. First, single cell clones derived from female mammals heterozygous at X-linked loci express one allele or the other, but never both. Second, all such single cell clones express only a single allele even after extensive cell division in vitro (Beutler et al., 1962; Davidson et al., 1963; Migeon et al., 1968; Chen et al., 1971; Goldstein et al., 1971; Gartler et al., 1972a; Chapman and Shows, 1976). These observations indicate that extensive reactivation of X linked loci on the inactive X does not occur either in vivo or in vitro. However, the techniques used in these studies to detect activity of loci on the inactive X, primarily gel electrophoresis followed by visualization with an enzyme activity stain, are relatively insensitive. Expression of the allele on the inactive X would have to contribute at least 5-10% of the enzyme activity detected.

Studies using a much more sensitive technique have confirmed the conclusion of these early studies. HPRT deficient single cell clones were isolated from a female human heterozygous for HPRT deficiency. These cells contain an active X chromosome with a deficient Hprt gene and an inactive X chromosome with a normal Hprt gene. These cells were grown in medium containing hypoxanthine, aminopterin and thymidine (HAT). Since cells must express HPRT in order to grow in HAT-containing medium, reactivation of the normal Hprt gene on the inactive X must occur for the cells to grow (assuming that the reversion frequency of the deficient Hprt gene on the active X is zero). Out of the approximately 10^7 cells plated in HAT-containing medium, no cells were able to grow, confirming and extending previous demonstrations of the stability of the inactive X (Migeon, 1972). Similar analyses of two HPRT-deficient fibroblast cell lines (B5TG) derived from a single cell clone from a *Mus musculus* x *Mus caroli* F-1 hybrid fetus (Chapman and Shows, 1976), also demonstrate the relative stability of the inactive X: reactivation rates of $\sim 10^{-8}$ and 10^{-6} were observed (Graves, 1982; Graves and Young, 1982; Lock and Martin, unpublished observation).

Thus, the fourth major point of the single active X hypothesis has been confirmed; repression of the genes on the inactive X is relatively stable and clonally inherited in normal diploid cells.

THE MECHANISM OF X INACTIVATION

The process of X inactivation can be considered as a two-step process. First, there is the primary inactivation event in which one of two active X chromosomes in the female embryo is inactivated. Second, there are events that result in the maintenance of repression of the genes on the inactivated X throughout numerous subsequent mitotic divisions and differentiative events. Although these two steps make up a single process, that of X inactivation, the mechanistic interrelationship between these two steps is unclear. Any model of X inactivation, in addition to being consistent with the four basic tenets of the single active X hypothesis, must explain both of these steps in the X inactivation process.

Over the years, numerous models of X inactivation have been proposed, each consistent, to a greater or lesser extent, with the four basic tenets of the single active X hypothesis and addressing the question of how the two steps in the X inactivation process are achieved. Most of the models propose that X inactivation is established and maintained by the differentiation of the two X chromosomes in the female embryo, and the basis of this differentiation falls into one of three categories. First, several models suggest that the two X chromosomes differ in protein content, either histone or nonhistone, and that this difference results in the observed difference in expression (Lyon, 1971; Holliday and Pugh, 1975). Second, a superstructural difference between the two X chromosomes has been proposed. One such model suggests differential gene superstructure (Cook, 1973), whereas, the other model suggests differential nuclear organization based on binding of only one of the two X chromosomes to a site on the nuclear envelope (Comings, 1968). Third, several models have been suggested in which the two chromosomes differ at the DNA level. This proposed difference between the two X chromosomes involves either integration of a mobile genetic element (Grumbach et al., 1963; Cooper, 1971) or modification of DNA (Holliday and Pugh, 1975; Riggs, 1975; Sager and Kitchin, 1975).

Unequivocal evidence in support of any one of these models is lacking. However, experimental observations bearing on both steps in the X inactivation process have been made. These observations suggest that a major controlling center on the X chromosome exists that has a role in the primary inactivation event in both random and preferential paternal X inactivation, and that DNA methylation has a role in maintaining the repression of genes on the inactive X.

Studies Aimed at Elucidating the Mechanism of the Primary Inactivation Event

Since the primary inactivation event occurs in the early embryo, study of this step of the X inactivation process is difficult. However, one approach to this problem that has proven useful is the study of instances in which the probability of inactivation of an X chromosome is altered and the result is primary nonrandom inactivation. Such studies have been concerned with the effect of allelic variation at the Xce locus and the phenomenon of preferential paternal inactivation in the extraembryonic tissues in mice.

It has been proposed that the X chromosome controlling element, Xce, is the major center on the X that controls inactivation (Cattanach et al., 1969). This proposal was based on two observations. First, as previously discussed, in tissues in which "random" inactivation occurs, allelic variants of Xce can affect, in *cis*, the probability of inactivation of a given X chromosome. Second, in one case in the mouse, in which an X/autosome translocation (Searle's translocation) is nonrandomly inactivated, the breakpoint of the translocation maps to the region of the Xce locus (Lyon et al., 1964). It has been proposed that the Xce genes on the two X chromosomes in females compete for a factor that either causes inactivation of an X or ensures continued activity of an X. Although this locus might represent the major inactivation controlling center, multiple, secondary inactivation centers have also been suggested, based on the observation that in many X/autosome translocations noncontiguous X chromosomal material can be coordinately inactivated (Latt et al., 1976; Takagi, 1980; Disteche et al., 1981). As yet, none of these secondary inactivation centers has been identified.

It has also been suggested that the Xce is the site of imprinting of the paternal X, which results in nonrandom inactivation of the paternally derived X chromosome in extraembryonic cell lineages. The concept of imprinting, defined by Crouse, (1960) as the molecular event that causes homologous chromosomes to behave differently within the same nucleus, has been invoked to explain the differential probabilities of inactivation of homologous maternal and paternal X chromosomes during preferential paternal X inactivation. It has been suggested that the paternal genome is imprinted either during spermatogenesis (Cooper, 1971) or in the oocyte during fertilization (Brown and Chandra, 1973; Chandra and Brown, 1975) and that this plays an important role in X inactivation. The suggestion that the Xce locus is the site of imprinting of the paternal genome is based on the following observation. An extreme allele of Xce, Xce^c, when carried on the paternally derived X chromosome, can significantly reduce preferential inactivation of the paternally derived X in extraembryonic tissues; see Table 6-1 (Rastan

Table 6-1. X Inactivation in Crosses of Different Xce Genotypes

Cross	Genotype of Female Embryo Xm/Xp	Percent Allocyclic Xp	
		Yolk Sac Endoderm	Yolk Sac Mesoderm
1	Xce^a/Xce^a	92.8	55.5
2	Xce^c/Xce^a	92.7	67.2*
3	Xce^c/Xce^c	89.8	52.0
4	Xce^a/Xce^c	78.0	52.7*

Note: The inheritance of Xce^c allele on the paternally derived X chromosome (Xp) reduces the extent of preferential inactivation of the Xp from approximately 90% (crosses 1 to 3) to 78% (cross 4) in the yolk sac endoderm. (After Rastan and Cattanach, 1983). The percent allocyclic Xp observed in the yolk sac mesoderm in Crosses 2 and 4 is not consistent with expectations from other studies. See Rastan and Cattanach (1983) for discussion.

and Cattanach, 1983). This result was explained by the following hypothesis. Imprinting of the Xce locus on the paternal X modifies it and affects its ability to compete for the proposed factor that controls X activity. In general, such a modification would result in a higher probability of the paternal X being inactivated, however, in the case of the Xce^c allele, some alteration has rendered this allele resistant to imprinting.

Since the phenomenon of preferential paternal X inactivation may provide insight into the mechanism of the primary event of X inactivation, much interest in the way in which it occurs has been generated. It has been speculated that there are two different ways in which imprinting can occur. One possibility is that the imprinting of the paternal genome is temporary, and that there is a single mechanism of X inactivation by which the imprinted (paternal) X chromosome is recognized and preferentially inactivated; in the absence of an imprinted X, inactivation is random. In the case of the mouse it has been suggested by Lyon (1977) and Monk (1978), that the marking mechanism that differentiates the paternal and maternal X chromosomes in early embryogenesis disappears prior to gastrulation. The consequences of this are that: (1) preferential paternal inactivation occurs in trophectoderm and primitive endoderm, the cell lineages that undergo X inactivation prior to gastrulation, and that (2) random inactivation occurs in primitive ectoderm, the cell lineage that undergoes X inactivation during gastrulation.

An alternative hypothesis is that imprinting of the paternal genome is relatively permanent, with preferential paternal and random X inactivation occurring by two different mechanisms, imprinting dependent and imprinting independent, respectively. In the simplest case, imprinting of the paternal X chromosome itself would result in inactivation in trophectoderm and primitive endoderm but not in primitive ectoderm. However, the activity of the X chromosome of paternal origin in XO mice is not consistent with such a mechanism (Frels and Chapman, 1979; Papaioannou et al., 1981).

Potentially, the EC *in vitro* model system of embryonic development provides an opportunity to determine whether imprinting of the paternal genome is temporary or permanent. Ideally, it should be possible using established EC cell lines heterozygous for an X-linked marker, to determine whether random or nonrandom X inactivation occurs when these cells form derivatives of extraembryonic and fetal lineages in vitro. If nonrandom, preferential paternal inactivation takes place when extraembryonic derivatives are formed by EC cells which have been maintained in the undifferentiated state for a large number of cell divisions, then the mechanism by which the maternal and paternal X chromosomes are distinguished is not lost due to temporal factors. In contrast, if random inactivation occurs in extraembryonic derivatives, then imprinting of the paternal genome may be temporary; imprinting having been lost during establishment of the cell line. For an EC cell line to be appropriate for such an analysis, it would have to be a pluripotent female cell line derived from a female embryo heterozygous for electrophoretic variants of an X-linked gene, such as Pgk-1. Such cell lines have been isolated, but stable retention of both X chromosomes in these cell lines appears to be rare, if possible at all, and the available cell lines have not proven useful for this experiment (McBurney and Strutt, 1980; Martin and Lock, 1983; Martin et al., unpublished observations).

Although there is as yet no direct evidence in support of either the "temporary" or the "permanent" hypothesis, two observations suggest that the mechanisms of nonrandom and random X inactivation differ. First, in tissues in which preferential paternal inactivation occurs, the inactive X chromosome replicates early in the synthetic phase of the cell cycle as compared with the late replication pattern seen in tissues in which random inactivation occurs (Takagi et al., 1978; 1982). Second, DNA isolated from the inactive X of yolk sac endoderm (nonrandom inactivation) functions in DNA-mediated transformation of HPRT deficient cells, whereas DNA isolated from the inactive X of yolk sac mesoderm (random inactivation) does not (Kratzer et al., 1983). The latter result, however, could be a consequence of differences not in the inactivation event but rather in the secondary or maintenance function.

Studies of the Mechanism by Which Repression of Genes on the Inactive X Chromosome Is Maintained

One approach that has been taken to elucidate the mechanism by which the inactivity of an entire X chromosome is stably maintained is to attempt to determine what conditions will lead to the reactivation of the inactive X. The numerous manipulations that have been tested for their ability to cause reactivation of the inactive X fall into two categories, somatic cell hybridization and chemical treatment.

Numerous studies have been conducted to determine whether fusion of a cell containing an inactive X with any of a variety of cell types, including both differentiated and embryonic cells, leads to reactivation. Since it had been observed that DNA and RNA synthesis could be induced in relatively differentiated cells by fusion with less differentiated cells of another species, in the first of such studies, a clone of fibroblasts expressing G6PD A was isolated from a human female heterozygous for electrophoretic variants of G6PD (AB) and fused with HPRT deficient mouse "transformed" (i.e., less differentiated) fibroblasts containing a single active X (Migeon, 1972). Hybrids were selected on the basis of their ability to grow in HAT medium. In this experiment the only cells that could survive the selective procedure were those in which the HPRT activity was provided by the human Hprt gene on the active X and the transformed phenotype was provided by the mouse cells. Reactivation was evaluated in the resultant hybrids by assays for the expression of G6PD B, the G6pd form carried on the inactive X. The fact that reactivation of the inactive X was not observed in this study is not surprising, since the frequency of reactivation would have to be quite high to be detected when only four hybrids are analyzed. A number of other fusion studies have been performed using HPRT deficient fibroblast clones derived from female humans heterozygous for HPRT deficiency, followed by selection in HAT medium. In contrast to the fusion study by Migeon described above, in these cases, since neither HPRT deficient parental line was able to grow in HAT containing medium, the only cells that could survive the selection procedure were those rare hybrids in which the inactive X was being expressed. In such experiments reactivation of the inactive X could be detected at frequencies of approximately 10^{-6}. That these hybrids express the human Hprt gene was demonstrated by electrophoretic analysis of hybrid clones. However, it was found that these hybrid clones do not express other human X linked genes (Kahan and DeMars, 1975; Hellkuhl and Grzeschik, 1978; Kahan, 1978; Kahan and DeMars, 1980; Mohandas et al., 1981). Thus, the reactivation observed in these studies of fusions between two fibroblast cell lines is localized to the region of the Hprt gene. That the reactivated Hprt gene is still associated with the inactive X chromosome has been demonstrated by observation of an allocyclic X chromosome in the hybrids expressing human HPRT which is lost when the hybrids are grown under conditions that select against cells which are expressing HPRT (Hellkuhl and Grzeschik, 1978). A similar frequency of reactivation was observed in a study in which an HPRT deficient fibroblast clone (B5TG, derived from a

Mus musculus x *Mus caroli* F-1 hybrid embryo and containing a *Mus caroli* inactive X with a normal Hprt gene was fused with either HPRT deficient mouse fibroblasts or Chinese hamster ovary cells (Graves and Young, 1982). In one of the studies in which HPRT deficient human cells containing a normal Hprt gene on the inactive X were fused with HPRT deficient mouse cells containing a single X, extensive analysis of the chromosomal constitution of the hybrids expressing human HPRT indicated that neither the active X chromosome or any single autosome is necessary to maintain inactivity, and that fragmentation of the inactive X does not necessarily lead to reactivation (Kahan and DeMars, 1980).

A second type of experiment involving somatic cell hybridization has been performed in an effort to reactivate the inactive X chromosome. These experiments were done with the idea that the pre-X inactivation environment of the EC cells, which are thought to be analogous to embryonic ectoderm prior to inactivation, contains factors that might cause derepression of the inactive X. In several studies fibroblasts containing an inactive X chromosome were fused with embryonal carcinoma (EC) cells. In one such study, B5TG cells were fused with HPRT deficient EC cells containing a single X (NG_2 or LT-TG EC cell lines) followed by selection in HAT containing medium (Graves and Young, 1982). Hybrids expressing the Hprt gene on the inactive X were isolated at a frequency of about 10^{-6}. Reactivation, however, was confined to only the Hprt gene. Also, the same reactivation frequency was obtained when hybrids were isolated following fusions of B5TG cells with fibroblasts. Thus, the frequency of reactivation observed in these EC/fibroblast hybrid studies did not differ significantly from that observed in the studies of hybrids between two fibroblast cell lines previously described. Recently, however, derepression of the entire inactive X chromosome has been demonstrated in hybrids derived from the fusion of mouse thymocytes with HPRT deficient EC cells (Takagi et al., 1983). In these hybrids, the previously inactive, allocyclic thymocyte X chromosome showed manifestations of complete reactivation. It replicated in synchrony with the autosomes and active X occurred in all hybrids analyzed. Thus, it appears that fusion can lead to not only localized reactivation of the Hprt gene but also to reactivation of the entire inactive X chromosome.

The difference in the extent of reactivation of the inactive X observed in these two studies involving fusions with EC cells may be related to the differentiated cell type used, and its effect on the phenotype of the hybrid. Hybrids between EC cells and thymocytes are usually EC-like by morphological criteria and are pluripotent (Miller and Ruddle, 1976, 1977), whereas hybrids between fibroblasts and EC cells are generally fibroblastlike in morphology (Finch and Ephrussi, 1967; McBurney and Strutt, 1979). The hybrids produced in the study by Takagi et al. (1983), by fusion of thymocytes and EC cells, were EC-like in morphology and were shown to be pluripotent. The hybrids produced in the study by Graves and Young (1982) by fusion of fibroblasts and EC cells were, surprisingly, of either fibroblast or EC-like morphology. It was not determined, however, whether the hybrids of EC-like morphology were pluripotent. It is possible that either the difference in pluripotency or some other undefined difference between the EC-like hybrids derived from fibroblast-EC cell fusions and EC-like hybrids derived from thymocyte-EC cell fusions accounts for the limited reactivation observed in the former as opposed to the dramatic reactivation of the entire X in the latter.

A second approach that has been taken to the study of the mechanism of repression of the genes on the inactive X has been to test the ability of various physical and chemical treatments to reactivate genes on the inactive X. Determination of the conditions under which the inactive X can be derepressed might provide insight into the mechanism by which the repression is maintained. In the first of such studies, a single cell clone derived from a female human heterozygous for electrophoretic variants of G6PD (A/B) was exposed to various physical and chemical treatments, including agents that increase RNA transcription. No evidence of reactivation was found. However, reactivation of the inactive X was assayed by tests for the expression of the G6pd allele on the inactive X (Comings, 1966b). As discussed previously, this is a relatively insensitive assay.

In more recent studies, localized reactivation of regions of the inactive X was observed at a high frequency (10^{-4}) after treatment of cells with chemical compounds known to result in hypomethylation of DNA (Mohandas et al., 1981; Jones et al, 1982; Graves, 1982; Lester et al., 1982). In the first of these studies, HPRT deficient human-mouse fusion hybrid cells containing an inactive human X, but no active human X, were treated with 5-azacytidine and selected in HAT-containing medium in which growth of the cells is dependent on reactivation of the Hprt gene on the inactive human X. Numerous colonies were isolated that expressed human HPRT. Reactivation of the unselected X-linked genes, human G6pd and Pgk-1, was observed in some reactivants; however, reactivation of the entire inactive X chromosome was not observed. These results were confirmed in several studies using either independently derived human-mouse fusion hybrid cells or an HPRT deficient clonal cell line derived from a *Mus musculus* x *Mus caroli* F-1 hybrid fetus (B5TG—see Table 6-2; Jones et al., 1982; Graves, 1982; Lester et al., 1982). Thus, it appears that 5-azacytidine treatment of cells can result in localized reactivation of regions of the inactive X chromosome. Presumably, this reactivation of genes on the inactive X results from demethylation of methylcytosine in a region(s) of DNA with regulatory importance in the repression of genes on the inactive X. This idea is based on

Table 6-2. Reactivation of X Linked Genes by Treatment with 5-azacytidine

Cell Type	Inactive X	Number of HAT Resistant Colonies Analyzed	Frequency of Coreactivation (%)			Reference
			G6PD	PGK	AGS	
fusion hybrid	human	14	7.1	7.1	n.a.	Mohandas et al., 1981
fusion hybrid	human	95	2.1	2.1	n.a.	Lester et al., 1982
fusion hybrid	human	62	12.9	4.8	62.9	Shapiro and Mohandas, 1983
fusion hybrid	human t(X/13)	14	7.1	0	n.a.	Mohandas et al., 1982
interspecific hybrid	*Mus caroli*	102	5.9	0	n.a.	Graves, 1982

Note: The frequency of coreactivation of the unselected X linked genes, G6PD, PGK, and AGS, in 5-acacytidine-induced reactivants expressing HPRT after HAT selection varies from 0 to 62.9% n.a., not analyzed.

the fact that 5-azacytidine treatment has resulted in hypomethylation of DNA in a number of systems (Christman et al., 1980; Jones and Taylor, 1980). Several observations support this idea. First, hypomethylation of DNA by treatment with 5-azacytidine, an analog of cytidine modified at the 5-position of the cytosine ring, is thought to result from replacement, at random, of 5-methylcytosine in the DNA by 5-azacytidine. Such substitution prevents the maintenance methylase from methylating the modified 5-position of 5-azacytidine (see Fig. 6-5). Several other cytidine analogs have been assayed for the ability to reactivate genes on the inactive X. Analogs modified at the 5-position can cause reactivation, whereas, analogs lacking modification at the 5-position do not (Jones et al., 1982; Lester et al., 1982). In some cases, hypomethylation of a specific gene has resulted in a change in expression of that gene (Niwa and Sugahara, 1981; Harris, 1982). Second, azacytidine treatment of mouse/human hybrid cells that results in reactivation of the Hprt gene on the inactive X also results in hypomethylation of DNA in these cells (Jones et al., 1982). Third, treatment with other chemicals that can cause hypomethylation of DNA, such as butyrate and dimethylsulfoxide, also causes reactivation of genes on the inactive X (Lester et al., 1982). These observations suggest that 5-azacytidine leads to hypomethylation of DNA.

The experiments described above are consistent with the proposal that DNA methylation is responsible for the heritable repression of genes on the inactive X (Riggs, 1975; Sager and Kitchin, 1975; Mohandas et al., 1981). An inverse correlation between cytosine methylation and expression of genes has been observed in a number of systems (Adams and Burdon, 1983). Such methylation patterns are replicated with high fidelity during DNA synthesis and can be perpetuated through many mitotic divisions (Adams and Burdon, 1983).

Two approaches have been taken to further explore the proposed role of DNA methylation in the repression of genes on the inactive X. First, the ability of DNAs isolated from active, inactive and 5-azacytidine reactivated X chromosomes have been assayed for their ability to transform HPRT deficient fibroblasts (Liskay and Evans, 1980; de Jonge et al., 1982; Venolia and Gartler, 1983). In the first of such studies, DNA was isolated from two cell lines: (1) a near-diploid female mouse cell line that contains two normal

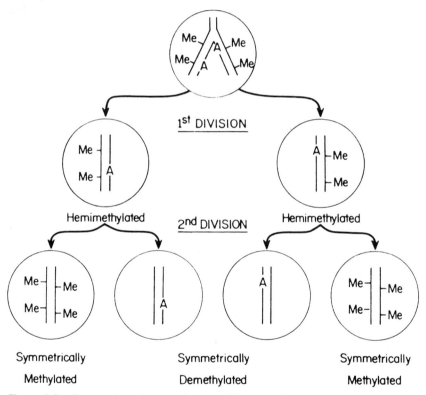

Figure 6-5. Proposed mechanism of action of 5-azacytidine.
The replication of X chromosome DNA containing 5-methylcytosine (ME) in the latter part of the S phase is shown at the top of the diagram. If 5-azacytosine (A) were incorporated into the newly synthesized DNA and inhibited DNA methylation, the daughter cells of the first division would contain hemimethylated DNA. Symmetrically demethylated DNA in this sequence would result after the second division after treatment, because of the specificity of the DNA methyltransferase. (From Jones et al., 1982)

X chromosomes (CAK-wt), and (2) an HPRT deficient subclone of CAK-wt isolated by treatment of CAK-wt cells with ethyl methanesulfonate followed by selection in HAT-containing medium (CAK-HP) (Liskay and Evans, 1980). About 65% of both CAK-wt and CAK-HP retain the inactive X chromosome. The DNA isolated from each of these cell lines was assayed for its ability to function in DNA-mediated transformation of HPRT deficient Chinese hamster cells in *in vitro*. DNA isolated from CAK-wt cells was able to transform at a frequency of about 2.5×10^{-7}, whereas, DNA from CAK-HP cells was not able to transform the HPRT deficient cells at all. DNA from each cell line was able to transform cells deficient at the autosomal gene, thymidine kinase, at roughly equal frequencies (3×10^{-7}). Since the normal Hprt gene is on the active X in CAK-wt and on the inactive X in CAK-HP, these data suggest that the Hprt genes on the active and inactive X chromosomes differ at the level of the DNA. Subsequent studies have shown that DNA isolated from the 5-azacytidine reactivated inactive X chromosome, like the DNA from the active X, will function in DNA-mediated transformation of the Hprt gene (Lester et al., 1982; Venolia et al., 1982).

A second approach has been taken to investigate the role of DNA methylation in the process of X inactivation. The methylation of specific sites in several X-linked genes on the active, inactive and reactivated X chromosomes has been compared (Wolf and Migeon, 1982; Wolf et al., 1984; Yen et al., 1984). In one of these studies, DNA was isolated from human/mouse hybrid cells containing either an active, inactive, or reactivated human X chromosome, then digested with methylation-sensitive restriction endonucleases, size fractionated by gel electrophoresis, transferred to nitrocellulose filters, and hybridized to DNA probes specific to the human Hprt gene (Yen et al., 1984). Comparison of the sizes of restriction fragments generated from the Hprt genes on the active, inactive or reactivated X chromosomes indicates that although no specific methylated sites are correlated with activity, an overall hypomethylation of the 5' region of the Hprt gene was correlated with activity. These studies, however, were concerned exclusively with restriction sites recognized by methylation sensitive endonucleases within approximately 5 to 15 kilobases of a 40 kilobase gene. It is conceivable that sites not examined in these studies are involved in the maintenance of repression of genes on the inactive X chromosome. Elucidation of the precise role of DNA methylation in the mechanism X inactivation requires further examination.

SUMMARY

The single active X hypothesis consists of four major tenets. First, there is only one actively expressed X chromosome in each cell of female mammals.

Second, inactivation of one of the two X chromosomes occurs during early embryogenesis. Third, the X chromosome of either maternal or paternal origin can become inactivated with the choice being made at random. Fourth, repression of genes on the inactive X is stable and clonally inherited. Evidence has been amassed in support of these tenets with only a few minor modifications. These include: (1) the lace of inactivation of some X-linked loci in some mammalian species, and (2) preferential inactivation of the paternally derived X chromosome in extraembryonic tissues of at least some mammalian species.

Elucidation of the mechanism of X inactivation has not been accomplished as yet. However, significant insights into the mechanism have been provided by several observations. Among these are: (1) the identification of a major inactivation controlling center that has a role in both random and nonrandom X inactivation; (2) the complete reactivation of the entire inactive X of thymocytes by fusion with pluripotent EC cells; and (3) the regional reactivation of the inactive X by hypomethylation caused by incorporation of 5-azacytidine. Further investigation should make it possible to conceive of a mechanism(s) of X inactivation that is consistent with these observations and the four basic tenets of the single active X hypothesis as proposed by Lyon (1961, 1962).

REFERENCES

Adams, R. L. P.; and Burdon, R. H. (1983). DNA methylation in eukaryotes. *CRC Crit. Rev. Biochem.* **13,** 349-384.

Albertini, R. J.; and DeMars, R. (1974). Mosaicism of peripheral blood lymphocyte populations in female heterozygous for the Lesch-Nyhan mutation. *Biochem. Genet.* **11,** 397-411.

Baverstock, P. R.; Adams, M.; Polkinghorne, R. W.; and Gelder, M. (1982). A sex-linked enzyme in birds—Z-chromosome conservation but no dosage compensation. *Nature* **296,** 763-766.

Beddington, R. (1983). The origin of foetal tissues during gastrulation in the rodent. In *Development in Mammals.* (M. H. Johnson, ed.), vol. 5, Elsevier, New York, pp. 3-32.

Beutler, E.; Yeh, M.; and Fairbanks, V. F. (1962). The normal human female as a mosaic of X-chromosome activity: Studies using the gene for G-6PD-deficiency as a marker. *Proc. Nat. Acad. Sci. (U.S.A.)* **48,** 9-16.

Brown, S. W.; and Chandra, H. S. (1973). Inactivation system of the mammalian X chromosome. *Proc. Natl. Acad. Sci. (U.S.A.)* **70,** 195-199.

Cattanach, B. M.; and Isaacson, J. H. (1967). Controlling elements in the mouse X chromosome. *Genetics* **57,** 331-346.

Cattanach, B. M.; Pollard, C. E.; and Perez, J. N. (1969). Controlling elements in the mouse X-chromosome. I. Interaction with the X-linked genes. *Genet. Res.* **14,** 223-235.

Cattanach, B. M.; Perez, J. N.; and Pollard, C. E. (1970). Controlling elements in the mouse X-chromosome. II. Location in the linkage map. *Genet. Res.* **15,** 183-195.

Chance, P. F.; and Gartler, S. M. (1983). Evidence for dosage effect at the X-linked steroid sulfatase locus in human tissues. *Am. J. Hum. Genet.* **35,** 234-240.

Chandra, H. S.; and Brown, S. W. (1975). Chromosome imprinting and the mammalian X chromosome. *Nature* **253,** 165-168.

Chapman, V. M.; and Shows, T. B. (1976). Somatic cell genetic evidence for X-chromosome linkage of three enzymes in the mouse. *Nature* **259**, 665-667.

Chen, S.-H.; Malcolm, L. A.; Yoshida, A.; and Giblett, E. R. (1971). Phosphoglycerate kinase: An X-linked polymorphism in man. *Am. J. Hum. Genet.* **23**, 87-91.

Christman, J. K.; Weich, N.; Schoenbrun, B.; Schneiderman, N.; and Acs, G. (1980). Hypomethylation of DNA during differentiation of Friend erythroleukemia cells. *J. Cell Biol.* **86**, 366-370.

Cohen, M. M.; and Rattazzi, M. C. (1971). Cytological and biochemical correlation of late X-chromosome replication and gene inactivation in the mule. *Proc. Nat. Acad. Sci. (U.S.A.)* **68**, 544-548.

Comings, D. E. (1966a). Uridine-5-H^3 radioautography of the human sex chromatin body. *J. Cell Biol.* **28**, 437-441.

Comings, D. E. (1966b). The inactive X chromosome. *Lancet* **ii**, 1137-1138.

Comings, D. E. (1968). The rationale for an ordered arrangement of chromatin in the interphase nucleus. *Am. J. Hum. Genet.* **20**, 440-460.

Cook, P. R. (1973). Hypothesis on differentiation and the inheritance of gene superstructure. *Nature* **245**, 23-26.

Cooper, D. W. (1971). Directed genetic change model for X chromosome inactivation in eutherian mammals. *Nature* **230**, 292-294.

Cooper, D. W.; VandeBerg, J. L.; Sharman, G. B.; and Poole, W. E. (1971). Phosphoglycerate kinase polymorphism in kangaroos provides further evidence for paternal X inactivation. *Nature New Biol.* **230**, 155-157.

Crocker, M.; and Craig, I. (1983). Variation in regulation of steroid sulphatase locus in mammals. *Nature* **303**, 721-722.

Crouse, H. V. (1960). The controlling element in sex chromosome behavior in *Sciara*. *Genetics* **45**, 1429-1443.

Davidson, R. G.; Nitowsky, H. M.; and Childs, B. (1963). Demonstration of two populations of cells in the human female heterozygous for glucose-6-phosphate dehydrogenase variants. *Proc. Natl. Acad. Sci. (U.S.A.)* **50**, 481-485.

de Jonge, A. J. R.; Abrahams, P. L.; Westerveld, A.; and Bootsma, D. (1982). Expression of human *hprt* gene on the inactive X chromosome after DNA-mediated gene transfer. *Nature* **295**, 624-626.

Deol, M. S.; and Whitten, W. K. (1972a). Time of X chromosome inactivation in retinal melanocytes of the mouse. *Nature New Biol.* **238**, 159-160.

Deol, M. S.; and Whitten, W. K. (1972b). X-chromosome inactivation: Does it occur at the same time in all cells of the embryo? *Nature New Biol.* **240**, 277-279.

Disteche, C. M.; Eicher, E. M.; and Latt, S. A. (1979). Late replication in an X-autosome translocation in the mouse: Correlation with genetic inactivation and evidence for selective effects during embryogenesis. *Proc. Natl. Acad. Sci. (U.S.A.)* **76**, 5234-5238.

Disteche, C. M.; Eicher, E. M.; and Latt, S. A. (1981). Late replication patterns in adult and embryonic mice carrying Searle's X-autosome translocation. *Exp. Cell Res.* **133**, 357-362.

Eicher, E. (1970). X-autosome translocations in mouse: total inactivation versus partial inactivation of the X chromosome. *Adv. Genet.* **15**, 175-259.

Epstein, C. J. (1969). Mammalian Oocytes: X chromosome activity. *Science* **163**, 1078-1079.

Epstein, C. J. (1972). Expression of the mammalian X chromosome before and after fertilization. *Science* **175**, 1467-1468.

Epstein, C. J.; Smith, S.; Travis, B.; and Tucker, G. (1978). Both X chromosomes function before visible X-chromosome inactivation female mouse embryos. *Nature* **274**, 500-502.

Evans, H. J.; Ford, C. E.; Lyon, M. F.; and Gray, J. (1965). DNA replication and genetic expression in female mice with morphologically distinguishable X chromosomes. *Nature* **206**, 900-903.

Felix, J. S.; and DeMars, R. (1969). Purine requirement of cells cultured from humans affected with Lesch-Nyhan syndrome (hypoxanthine-guanine phosphoribosyltransferase deficiency). *Proc. Natl. Acad. Sci. (U.S.A.)* **62**, 536-543.

Fialkow, P. J. (1970). X-chromosome inactivation and the Xg locus. *Am. J. Hum. Genet.* **22**, 460-463.

Fialkow, P. J.; Lisker, R.; Giblett, E. R.; and Zavala, C. (1970). Xg locus: failure to detect inactivation in females with chronic myelocytic leukaemia. *Nature* **266**, 367-368.

Finch, B. W.; and Ephrussi, B. (1967). Retention of multiple developmental potentialities by cells of a mouse testicular teratocarcinoma during prolonged culture *in vitro* and their extinction upon hybridization with cells of permanent lines. *Proc. Natl. Acad. Sci. (U.S.A.)* **57**, 615-621

Frels, W. I.; and Chapman, V. M. (1979). Paternal X chromosome expression in extraembryonic membranes of XO mice. *J. Exp. Zool.* **210**, 553-560.

Frels, W. I.; and Chapman, V. M. (1980). Expression of the maternally derived X chromosome in the mural trophoblast of the mouse. *J. Embryol. Exp. Morph.* **56**, 179-190.

Frels, W. I.; Rossant, J.; and Chapman, V. M. (1979). Maternal X chromosome expression in mouse chorionic ectoderm. *Devel. Genet.* **1**, 123-132.

Gandini, E.; Gartler, S. M.; Angioni, G.; Argiolas, N.; and Dell'Acqua, G. (1968). Developmental implication of multiple tissue studies in glucose-6-phosphate dehydrogenase deficient heterozygotes. *Proc. Natl. Acad. Sci. (U.S.A.)* **61**, 945-948.

Gardner, R. L.; and Lyon, M. F. (1971). X chromosome inactivation studied by injection of a single cell into the mouse blastocyst. *Nature* **231**, 385-386.

Gardner, R. L.; Papaioannou, V. E. (1975). Differentiation in the trophectoderm and inner cell mass. In *The Early Development of Mammals* (M. Balls and A. E. Wild, eds.), Cambridge University Press, Cambridge. pp. 107-132.

Gartler, S. M. (1976). X-chromosome inactivation and selection in somatic cells. *Fed. Proc.* **35**, 2191-2194.

Gartler, S. M.; and Rivest, M. (1983). Evidence for X-linkage of steroid sulfatase in the mouse: steroid sulfatase levels in oocytes of XX and XO mice. *Genetics* **103**, 137-141.

Gartler, S. M.; and Sparkes, R. S. (1963). The Lyon-Beutler hypothesis and isochromosome X patients with Turner Symdrome. *Lancet* **2**, 411.

Gartler, S. M.; Andina, R.; and Gant, N. (1975). Ontogeny of X-chromosome inactivation in the female germ line. *Exp. Cell. Res.* **91**, 454-457.

Gartler, S. M.; Chen, S.-H.; Fialkow, P. J.; Giblett, E. R.; and Singh, S. (1972a). X chromosome inactivation in cells from an individual heterozygous for two X-linked genes. *Nature New Biol.* **236**, 149-150.

Gartler, S. M.; Liskay, R. M.; Campbell, B. K.; Sparkes, R.; and Gant, N. (1972b). Evidence for two functional X chromosomes in human oocytes. *Cell Diff.* **1**, 215-218.

Gartler, S. M.; Liskay, R. M.; and Gant, N. (1973). Two functional chromosomes in human fetal oocytes. *Exp. Cell Res.* **82**, 464-466.

Gartler, S. M.; Rivest, M.; and Cole, R. E. (1980). Cytological evidence for an inactive X chromosome in murine oogonia. *Cytogenet. Cell Genet.* **28**, 203-207.

Giannelli, F.; and Hamerton, J. L. (1971). Non-random late replication of X chromosomes in mules and hinnies. *Nature* **232**, 315-319.

Goldstein, J. L.; Marks, J. F.; and Gartler, S. M. (1971). Expression of two X-linked genes in human hair follicles of double heterozygotes. *Proc. Natl. Acad. Sci. (U.S.A.)* **68**, 1425-1427.

Graves, J. A. M. (1982). 5-azacytidine-induced re-expression of alleles on the inactive X chromosome in a hybrid mouse cell line. *Exp. Cell Res.* **141**, 99-105.

Graves, J. A. M.; and Young, G. J. (1982). X-chromosome activity in heterokaryons and hybrids between mouse fibroblasts and teratocarcinoma stem cells. *Exp. Cell Res.* **141**, 87-97.

Grumbach, M. M.; Morishima, A.; and Taylor, J. H. (1963). Human sex chromosome abnormalities in relation to DNA replication and heterochromatinization. *Proc. Natl. Acad. Sci. (U.S.A.)* **49**, 581-589.

Hamerton, J. L.; Richardson, B. J.; and Gee, P. A. (1971). Non-random X chromosome expression in female mules and hinnies. *Nature* **232,** 312-314.

Harper, M. I.; Fosten, M.; and Monk, M. (1982). Preferential paternal X inactivation in extraembryonic tissues of early mouse embryos. *J. Embryol. Exp. Morph.* **67,** 127-135.

Harris, M. (1982). Induction of thymidine kinase in enzyme-deficient Chinese hamster ovary cells. *Cell,* **29** 483-492.

Hellkuhl, B.; and Grzeschik, K.-H. (1978). Partial reactivation of human inactive X chromosome in human-mouse somatic cell hybrids. *Cytogenet. Cell Genet.* **22,** 527-530.

Holliday, R.; and Pugh, J. E. (1975). DNA modification mechanisms and gene activity during development. *Science* **187,** 226-232.

Hook, E. B.; and Brustman, L. D. (1971). Evidence for selective differences between cells with an active horse X chromosome and cells with an active donkey X chromosome in the female mule. *Nature* **232,** 349-350.

Issa, M.; Blank, C. E.; and Atherton, G. W. (1969). The temporal appearance of sex chromatin and the late-replicating X chromosome in blastocysts of the domestic rabbit. *Cytogenetics* **8,** 219-237.

Johnson, M. S.; and Turner, J. R. G. (1979). Absence of dosage compensation for a sex-linked enzyme in butterflies (Heliconius). *Heredity* **43,** 71-77.

Johnston, P. G. (1981). X chromosome activity in female germ cells of mice heterozygous for Searle's translkocation T(X;16)-16H. *Genet. Res., Camb.* **37,** 317-322.

Johnston, P. G.; and Cattanach, B. M. (1981). Controlling elements in the mouse. IV. Evidence of non-random X-inactivation. *Genet. Res.* **37,** 151-160.

Johnston, P. G.; Sharman, G. B.; James, E. A.; and Cooper, D. W. (1978). Studies on metatherian sex chromosomes. VII. Glucose-6-phosphate dehydrogenase expression in tissues and cultured fibroblasts of kangaroos. *Aust. J. Biol. Sci.* **31,** 415-424.

Jones, P. A.; and Taylor, S. M. (1980). Cellular differentiation, cytidine analogs and DNA methylation. *Cell* **20,** 85-93.

Jones, P. A.; Taylor, S. M.; Mohandas, T.; and Shapiro, L. J. (1982). Cell cycle-specific reactivation of an inactive X-chromosome locus by 5-azadeoxycytidine. *Proc. Natl. Acad. Sci. (U.S.A.)* **79,** 1215-1219.

Kahan, B. (1978). The stability of X-chromosome inactivation: Studies with mouse-human cell hybrids and mouse teratocarcinomas. In *Genetic Mosaics and Chimeras in Mammals* (L. B. Russell, ed.), pp. 297-328, Plenum, New York.

Kahan, B.; and DeMars, R. (1975). Localized derepression on the human inactive X chromosome in mouse-human cell hybrids. *Proc. Natl. Acad. Sci. (U.S.A.)* **72,** 1510-1514.

Kahan, B.; and DeMars, R. (1980). Autonomous gene expression on the inactive X chromosome. *Somat. Cell Genet.* **6,** 309-323.

Kazazian, H. H.; Young, W. J.; and Childs, B. (1965). X-linked 6-phosphogluconate dehydrogenase in *Drosophila*: Subunit associations. *Science* **150,** 1601-1602.

Keitges, E.; Rivest, M.; Siniscalco, M.; and Gartler, S. M. (1985). X-linkage of steroid sulphatase in the mouse is evidence for a functional Y-linked allele. *Nature* **315,** 226-227.

Kinsey, J. D. (1967). X-chromosome replication in early rabbit embryos. *Genetics* **55,** 337-343.

Kozak, L. P.; and Quinn, P. J. (1975). Evidence for dosage compensation of an X-linked gene in the 6-day embryo of the mouse. *Devel. Biol.* **45,** 65-73.

Kozak, L. P.; McLean, G. K.; and Eicher, E. M. (1974). X linkage of phosphoglycerate kinase in the mouse. *Biochem. Genet.* **11,** 41-47.

Kratzer, P. G.; and Chapman, V. M. (1981). X chromosome reactivation in oocytes of *Mus caroli. Proc. Natl. Acad. Sci. (U.S.A.)* **78,** 3093-3097.

Kratzer, P. G.; and Gartler, S. M. (1978). HGPRT activity changes in preimplantation mouse embryos. *Nature* **274,** 503-504.

Krazter, P. G.; Chapman, V. M.; Lambert, H.; Evans, R. E.; and Liskay, R. M. (1983). Differences in the DNA of the inactive X chromosomes of fetal and extraembryonic tissues of mice. *Cell* **33,** 37-42.

Latt, S. A.; Willard, H. F.; and Gerald, P. S. (1976). BrdU-33258 Hoechst analysis of DNA replication in human lymphocytes with supernumerary or structurally abnormal X chromosomes. *Chromosoma* **57**, 135-153.

Lester, S. C.; Korn, N. J.; and DeMars, R. (1982). Derepression of genes on the human inactive X chromosome: Evidence for differences in locus-specific rates of derepression and rates of transfer of active and inactive genes after DNA-mediated transformation. *Somat. Cell Genet.* **8**, 265-284.

Liskay, R. M.; and Evans, R. J. (1980). Inactive X chromosome DNA does not function in DNA-mediated cell transformation for the hypoxanthine phosphoribosyltransferase gene. *Proc. Natl. Acad. Sci. (U.S.A.)* **77**, 4895-4898.

Lyon, M. F. (1961). Gene action in the X-chromosome of the mouse (*Mus musculus* L.). *Nature* **190**, 372-373.

Lyon, M. F. (1962). Sex chromatin and gene action in the mammalian X-chromosome. *Am. J. Hum. Genet.* **14**, 135-148.

Lyon, M. F. (1971). Possible mechanisms of X chromosome inactivation. *Nature New Biol.* **232**, 229-232.

Lyon, M. F. (1977). Section II-Chairman's address. In *Reproduction and Evolution*, pp. 95-98. Fourth Intern. Symp. Comp. Biol. Reprod. Australian Academy of Science. Canberra.

McBurney, M. W.; and Strutt, B. J. (1979). Fusion of embryonal carcinoma cells to fibroblast cells, cytoplasts, and karyoplasts. *Exp. Cell Res.* **124**, 171-180.

McLaren, A.; Gauld, I. K.; and Bowman, P. (1973). Comparison between mice chiameric and heterozygous for the X-linked gene *tabby*. *Nature* **241**, 180-183.

McMahon, A.; Fosten, M.; and Monk, M. (1981). Random X-chromosome inactivation in female primordial germ cells in the mouse. *J. Embryol. Exp. Morph.* **64**, 251-258.

Mangia, F.; Abbo-Halbasch, G.; and Epstein, C. J. (1975). X chromosome expression during oogenesis in the mouse. *Devel. Biol.* **45**, 366-368.

Martin, G. R. (1980). Terectocarcinomas and mammalian embryogenesis. *Science* **209**, 768-776.

Martin, G. R.; and Lock, L. F. (1983). Pluripotent cell lines derived directly from early mouse embryos cultured in medium conditioned by teratocarcinoma stem cell. In *Teratocarcinoma Stem Cells* (L. M. Silver, G. R. Martin, and S. Strickland, eds.), vol. 10, pp. 635-646, Cold Spring Harbor Laboratory, Cold Spring Harbor, New York.

Martin, G. R.; Epstein, C. J.; Travis, B.; Tucker, G.; Yatziv, S.; Martin, D. W.; Clift, S.; and Cohen, S. (1978). X-chromosome inactivation during differentiation of female teratocarcinoma stem cells *in vitro*. *Nature* **271**, 329-333.

Meneely, P. M.; and Wood, W. B. (1984). An autosomal gene that affects X chromosome expression and sex determination in *Caenorhabditis elegans*. *Genetics* **106**, 29-44.

Migeon, B. R. (1970). Studies of skin fibroblasts from 10 families with HGPRT deficiency, with reference to X-chromosomal inactivation. *Am. J. Hum. Genet.* **23**, 199-210.

Migeon, B. R. (1972). Stability of X chromosomal inactivation in human somatic cells. *Nature* **239**, 87-89.

Migeon, B. R. (1978). Selection and cell communication as determinants of female phenotype. In *Genetic Mosaics and Chimeras in Mammals*. (L. B. Russell, ed.) pp. 417-444, Plenum, New York.

Migeon, B. R.; and Jelalian, K. (1977). Evidence for two active X chromosomes in germ cells of female before meiotic entry. *Nature* **269**, 242-243.

Migeon, B. R.; Der Kaloustian, V. M.; Nyhan, W. L.; Young, W. J.; Childs, B. (1968). X-linked hypoxanthine-guanine phosphoribosyl transferase deficiency: Heterozygote has two clonal populations. *Science* **160**, 425-427.

Migeon, B. R.; Shapiro, L. J.; Norum, R. A.; Mohandas, T.; Axelman, J.; and Dabora, R. L. (1982). Differential expression of steroid sulphatase on active and inactive human X chromosome. *Nature* **299**, 838-840.

Miller, R. A.; and Ruddle, F. (1976). Pluripotent teratocarcinoma-thymus somatic cell hybrids. *Cell* **9**, 45-55.

Miller, R. A.; and Ruddle, F. (1977). Properties of teratocarcinoma-thymus somatic cell hybrids. *Somat. Cell Genet.* **3**, 247-261.

Mohandas, T.; Shapiro, L. J.; Sparkes, R. S.; and Sparkes, M. C. (1979). Regional assignment of the steroid sulfatase-X-linked ichthyosis locus: Implications for a noninactivated region on the short arm of human X chromosome. *Proc. Natl. Acad. Sci. (U.S.A.)* **76**, 5779-5783.

Mohandas, T.; Sparkes, R. S.; Hellkuhl, B.; Grzeschik, K. H.; and Shapiro, L. J. (1980). Expression of an X-linked gene from an inactive human X chromosome in mouse-human hybrid cells: Further evidence for the noninactivation of the steroid sulfatase locus in man. *Proc. Natl. Acad. Sci. (U.S.A.)* **77**, 6759-6763.

Mohandas, T.; Sparkes, R. S.; and Shapiro, L. J. (1981). Reactivation of an inactive human X chromosome: Evidence for X inactivation by DNA methylation. *Science* **211**, 393-396.

Monk, M. (1978). Biochemical studies on mammalian X-chromosome activity. In *Development in Mammals* (M. H. Johnson, ed.), vol. 3, pp. 189-223, Elsevier, Amsterdam.

Monk, M.; and Kathuria, H. (1977). Dosage compensation for an X-linked gene in preimplantation mouse embryos. *Nature* **270**, 599-601.

Monk, M.; and Harper, M. I. (1978). X-chromosome activity in preimplantation mouse embryos from XX and XO mothers. *J. Embryol. Exp. Morph.* **46**, 53-64.

Monk, M.; and Harper, M. I. (1979). Sequential X chromosome inactivation coupled with cellular differentiation in early mouse embryos. *Nature* **281**, 311-313.

Monk, M.; and McLaren, A. (1981). X-chromosome activity in foetal germ cells of the mouse. *J. Embryol. Exp. Morph.* **63**, 75-84.

Morishima, A.; Grumbach, M. M.; and Taylor, J. H. (1962). Asynchronous duplication of human chromosomes and the origin of sex chromatin. *Proc. Natl. Acad. Sci. (U.S.A.)* **48**, 756-763.

Mukerhjee, A. S.; and Beermann, W. (1965). Synthesis of ribonucleic acid by the X-chromosomes of *Drosophila melanogaster* and the problem of dosage compensation. *Nature* **207**, 785-786.

Mukherjee, B. B.; and Milet, R. G. (1972). Nonrandom X-chromosome inactivation — an artifact of cell selection. *Proc. Natl. Acad. Sci. (U.S.A.)* **69**, 37-39.

Mukherjee, B. B.; and Sinha, A. K. (1964). Single-active X hypothesis: Cytological evidence for random inactivation of X-chromosomes in a female mule complement. *Proc. Natl. Acad. Sci. (U.S.A.)* **51**, 252-259.

Muller, C. R.; Wahlstrom, J.; and Ropers, H.-H. (1981). Further evidence for the assignment of the steroid sulfatase X-linked ichthyosis locus to the telomer of Xp. *Hum. Genet.* **58**, 446.

Muller, H. J. (1950). Evidence of the precision of genetic adaption. *Harvey Lect.* **43**, 165-229.

Niwa, O.; and Sugahara, T. (1981). 5-azacytidine inducation of mouse endogenous type C virus and suppression of DNA methylation. *Proc. Natl. Acad. Sci. (U.S.A.)* **78**, 6290-6294.

Nyhan, W. L.; Bakay, B.; Connor, J. D.; Marks, J. F.; and Keele, D. K. (1970). Hemizygous expression of glucose-6-phosphate dehydrogenase in erythrocytes of heterozygotes for the Lesch-Nyhan syndrome. *Proc. Natl. Acad. Sci. (U.S.A.)* **65**, 214-218.

Ohno, S. (1967). *Sex chromosomes and sex linked genes.* Springer-Verlag, New York.

Ohno, S.; and Hauschka, T. S. (1960). Allocycly of the X-chromosome in tumors and normal tissues. *Cancer Res.* **20**, 541-545.

Ohno, S.; Kaplan, W. D.; and Kinosita, R. (1959). Formation of the sex chromatin by a single X-chromosome in liver cells of *Rattus norvegicus*. *Exp. Cell Res.* **18**, 415-418.

Ohno, S.; Poole, J.; and Gustavsson, I. (1965). Sex-linkage of erthrocyte glucose-6-phosphate dehydrogenase in two species of wild hare. *Science* **150**, 1737-1738.

Papaioannou, V. E.; and West, J. D. (1981). Relationship between the parental origin of the X chromosomes, embryonic cell lineage and X chromosome expression in mice. *Genet. Res.* **37**, 183-197.

Papaioannou, V. E.; West, J. D.; Bucher, T.; and Linke, I. M. (1981). Non-random X-chromosome expression early in mouse development. *Devel. Genet.* **2**, 305-315.

Race, R. R.; and Sanger, R. (1975). *Blood groups in man.* F. A. Davis, Philadelphia.
Rastan, S. (1982). Primary non-random X-inactivation caused by controlling elements in the mouse demonstrated at the cellular level. *Genet. Res.* **40,** 139-147.
Rastan, S.; and Cattanach, B. M. (1983). Interaction between the *Xce* locus and imprinting of the paternal X chromosome in mouse yolk-sac endoderm. *Nature* **303,** 635-637.
Rattazzi, M. C.; and Cohen, M. M. (1972). Further proof of genetic inactivation of the X chromosome in the female mule. *Nature* **237,** 393-396.
Ray, M.; Gee, P. A.; Richardson, B. J.; and Hamerton, J. L. (1972). G6PD expression and X chromosome late replication in fibroblast clones from a female mule. *Nature* **237,** 396-397.
Riggs, A. D. (1975). X inactivation, differentiation and DNA methylation. *Cytogenet. Cell Genet.* **14,** 9-15.
Ropers, H.-H.; and Wiberg, U. (1982). Evidence for X-linkage and non-inactivation of steroid sulphatase locus in wood lemming. *Nature* **296,** 766-767.
Ropers, H.-H.; Migl, B.; Zimmer, J.; Fraccaro, M.; Maraschio, P. P.; and Westerveld, A. (1981). Activity of steroid sulfatase in fibroblasts with numerical and structural X chromosome aberrations. *Hum. Genet.* **57,** 354-356.
Russell, L. B. (1961). Genetics of mammalian sex chromosomes. *Science* **133,** 1795-1803.
Russell, L. B. (1963). Mammalian X-chromosome action: Inactivation limited in spread and in region of origin. *Science* **140,** 976-978.
Russell, L. B.; and Cachiero, L. A. (1978). The use of X/autosome translocations in the study of X-inactivation pathways and nonrandomness. In *Genetic Mosaics and Chimeras in Mammals.* (L. B. Russell, ed.), pp. 393-416, Plenum, New York.
Sager, R.; and Kitchin, R. (1975). Selective silencing of dukaryotic DNA. *Science* **189,** 426-433.
Salzman, J.; DeMars, R.; and Benke, P. (1968). Single-allele expression at an X-linked hyperuricemia locus in heterozygous human cells. *Proc. Natl. Acad. Sci. (U.S.A.)* **60,** 545-552.
Searle, A. G. (1962). Is sex-linked *Tabby* really recessive in the mouse? *Heredity* **17,** 297.
Seecof, R. L.; Kaplan, W. D.; and Futch, D. G. (1969). Dosage compensation for enzyme activities in *Drosophila melanogaster. Proc. Natl. Acad. Sci. (U.S.A.)* **62,** 528-535.
Serov, O. L.; Zakijian, S. M.; and Kulichkov, V. A. (1978). Allelic expression in intergeneric fox hybrids (*Alopex lagopus* x *Vulpes vulpes*). III. Regulation of the expression of the parental alleles at the *Gpd* locus linked to the X chromosome. *Biochem. Genet.* **16,** 145-157.
Shapiro, L. J.; and Mohandas, T. (1983). DNA methylation and the control of gene expression on the human X chromosome. *Cold Spring Harbor Symp. Quant. Biol.* **47,** 631-639.
Shapiro, L. J.; Mohandas, T.; Weiss, R.; and Romeo, G. (1979). Non-inactivation of an X-chromosome locus in man. *Science* **204,** 1224-1226.
Shapiro, L. J.; Mohandas, T.; and Rotter, J. I. (1982). Dosage compensation at the steroid sulfatase (STS) locus. *Clin. Res.* **30,** 119A.
Sharman, G. B. (1971). Late replication in the paternally derived X chromosome of female kangaroos. *Nature* **230,** 231-232.
Takagi, N. (1974). Differentiation of X chromosomes in early female mouse embryos. *Exp. Cell Res.* **86,** 127-135.
Takagi, N. (1978). Preferential inactivation of the paternally derived X chromosome in mice. In *Genetic Mosaics and Chimeras in Mammals* (L. B. Russell, ed.), pp. 341-360, Plenum, New York.
Takagi, N. (1980). Primary and secondary nonrandom X chromosome inactivation in early female mouse embryos carrying Searle's translocation T(X;16)16H. *Chromosoma* **81,** 439-459.
Takagi, N.; and Martin, G. R. (1984). Studies of the temporal relationship between the cytogenetic and biochemical manifestations of X-chromosome inactivation during the differentiation of LT-1 teratocarcinoma stem cells. *Devel. Biol.* **103,** 425-433.
Takagi, N.; and Sasaki, M. (1975). Preferential inactivation of the paternally derived X chromosome in the extraembryonic membranes of the mouse. *Nature* **256,** 640-642.

Takagi, N.; Wake, N.; and Sasaki, M. (1978). Cytological evidence for preferential inactivation of the paternally derived X chromosome in XX mouse blastocysts. *Cytogenet. Cell Genet.* **20**, 240-248.

Takagi, N.; Sugawara, O.; and Sasaki, M. (1982). Regional and temporal changes in the pattern of X-chromosome replication during the early post-implantation development of the female mouse. *Chromosoma* **85**, 275-286.

Takagi, N.; Yoshida, M. A.; Sugawara, O.; and Sasaki, M. (1983). Reversal of X-inactivation in female mouse somatic cells hybridized with murine teratocarcinoma stem cells *in vitro*. *Cell* **34**, 1053-1062.

Vandeberg, J. L.; Cooper, D. W.; and Sharman, G. B. (1973). Phosphoglycerate kinase A polymorphism in the wallaby *Macropus parryi*: Activity of both X chromosomes in muscle. *Nature New Biol.* **243**, 47-48.

Venolia, L.; and Gartler, S. M. (1983). Comparison of transformation efficiency of human active and inactive X-chromosomal DNA. *Nature* **302**, 82-83.

Venolia, L.; Gartler, S. M.; Wassman, E. R.; Yen, P.; Mohandas, T.; and Shapiro, L. J. (1982). Transformation with DNA from 5-azacytidine-reactivated X chromosomes. *Proc. Natl. Acad. Sci., (U.S.A.)* **79**, 2352-2354.

Wake, N.; Takagi, N.; and Sasaki, M. (1976). Non-random inactivation of X chromosome in the rat yolk sac. *Nature* **262**, 580-581.

West, J. D.; and Chapman, V. M. (1978). Variation for X chromosome expression in mice detected by electrophoresis of phosphoglycerate kinase. *Genet. Res.* **32**, 91-102.

West, J. D.; Frels, W. I.; and Chapman, V. M. (1977). Preferential expression of maternally derived X chromosome in the mouse yolk sac. *Cell* **12**, 873-882.

Wolf, S. F.; and Migeon, B. R. (1982). Studies of X chromosome DNA methylation in normal human cells. *Nature* **285**, 667-671.

Wolf, S. F.; Jolly, D. J.; Lunnen, K. D.; Friedmann, T.; and Migeon, B. R. (1984). Methylation of the hypoxanthine phosphoribosyltransferase locus on the human X chromosome: Implications for X-chromosome inactivation. *Proc. Natl. Acad. Sci. (U.S.A.)* **81**, 2806-2810.

Yen, P. H.; Patel, P.; Chinault, A. C.; Mohandas, T.; and Shapiro, L. J. (1984). Differential methylation of hypoxanthine phosphoribosyltransferase genes on active and inactive human X chromosomes. *Proc. Natl. Acad. Sci. (U.S.A.)* **81**, 1759-1763.

7

Polytene Chromosomes

Milan Jamrich
*National Institute of Child Health
and Human Development
Bethesda, Maryland*

Polytene chromosomes were first identified as structural entities in 1881 in *Chironomus* salivary glands by Balbiani. It was Rambousek (1912) who recognized that these banded structures represent chromosomes. This important conclusion went unnoticed for many years, mainly because it was published in Czech. The rediscovery and analysis of these structures in the early thirties (Heitz and Bauer, 1933; Painter, 1933; King and Beams, 1934; Bauer, 1935; Caspersson, 1940) confirmed that they indeed represent giant interphase chromosomes. In the following five decades polytene chromosomes have proved to be an extremely useful tool for the analysis of DNA organization and gene expression.

POLYTENE NUCLEI

Polytene chromosomes can be found in a variety of protists, plants, and lower and higher animals (for review see Nagl, 1978). Characteristically, they are present in terminally differentiated cells which do not undergo any further cell division. They are a result of endomitosis, a process in which DNA is replicated without the separation of daughter chromatids. The daughter chromatids remain aligned side by side, increasing gradually the diameter of the chromosomes. Depending on the number of replication rounds, the size of these chromosomes can vary from tissue to tissue and cell to cell. In the polytene chromosomes of *Drosophila* salivary glands the chromosomes contain about 1,000 DNA fibers, in *Chironomus* about 16,000. They can be hundreds of μm long and 25 μm in diameter. Because of their proportions they are also called giant chromosomes. Their size is reflected in the size of the nuclei, since the diploid nuclei are not large enough to accommodate the additional DNA. The correlation between the size of nuclei and the degree of polyteny is demonstrated in Figure 7-1, which shows a portion of a *Drosophila* salivary gland stained with ethidium bromide. The large saliva producing cells have a higher polyteny than the excretory duct cells. Both cell types are larger in cell size than the diploid cells of the imaginal discs. The largest polytene nuclei in *Drosophila* are about 50 μm in diameter, about three times smaller than the largest known polytene nuclei of *Cryptochironomus* (Beermann, 1962). Figure 7-2 shows a polytene nucleus of *Chironomus* as seen through Nomarski optics. The nucleus is round in shape, a property characteristic of polytene nuclei (see Beermann, 1962). The location of the polytene chromosomes in the nucleus is not random. The chromosomes are positioned close to the nuclear envelope, leaving the center of the nucleus free. Certain chromosomal sites, such as the chromocenter and intercalary heterochromatin, seem to have firm attachment points on the nuclear envelope (Agard and Sedat, 1983; Mathog et al., 1984).

Fixation, staining, and mild squashing of polytene cells results in chromosome preparations such as those in Figure 7-3. The most obvious feature of these chromosomes is their typical banding pattern, which is shown in more detail in Figure 7-4a. In *Drosophila melanogaster* about 5,000 darkly staining bands alternate with lightly staining regions called interbands. The centromeres of all four chromosomes usually aggregate to form a structure with less prominent banding called the chromocenter. As seen in Figure 7-5, the chromocenter of *Drosophila* can be subdivided into two areas (Heitz, 1934). The darkly staining dot in the center is called the α-heterochromatin. This region consists of highly condensed underreplicated sequences, and is characterized by an absence of transcription (Berendes and Keyl, 1967; Rudkin, 1969; Gall et al., 1971). In males it also contains the entire Y-chromosome, which carries genes affecting male fertility. The more diffuse chromocentric region separating the α-heterochromatin from the chromosome arms is termed β-heterochromatin. This chromatin does contain genes (for review see Hilliker et al., 1980a) and is known to be transcribed in polytene cells (Lakhotia, 1974; Lakhotia and Jacob, 1974). It appears that only the α-heterochromatin is underreplicated. It is true that the banding pattern in β-heterochromatin is not easy to demonstrate, but this is likely to be due to the influence of α-heterochromatin. The β-heterochromatin shares common repetitive sequences with the telomeres (Young et al., 1983; Renkawitz-Pohl and Bialojan, 1984). Telomeres (Muller, 1938) are free chromosome ends without identified function. They frequently exhibit telomere-to-telomere

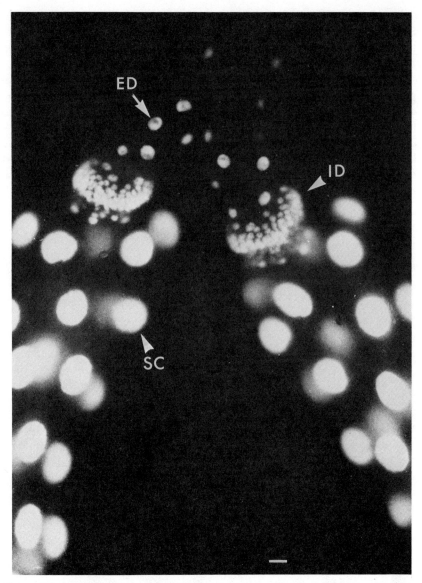

Figure 7-1. Salivary glands of *Drosophila melanogaster* stained with ethidium bromide. Note the size difference between the diploid cells of imaginal discs (ID), the highly polytene secretory cells (SC) and the intermediate cells of the excretory duct (ED). Bar represents 10 μm.

and telomere-to-nuclear membrane associations (Hughes-Schrader, 1943; Hinton, 1945). Occasionally associations between telomeres and centromeres can be observed, and at least in one case, in rye, the telomeres seem to be able to mimic centromeric functions (Ostergren and Praaken, 1946).

For decades the relatively simple squashing method for preparation of polytene chromosomes and their subsequent analysis by light or electron microscopy was the sole source of information about their structure and function. Although this technique, in combination with advances in *Drosophila* genetics, provided a constant flow of information, a detailed analysis of the DNA and protein distribution on different regions of the chromosomes was not possible. In the last 15 years three major technical improvements revolu-

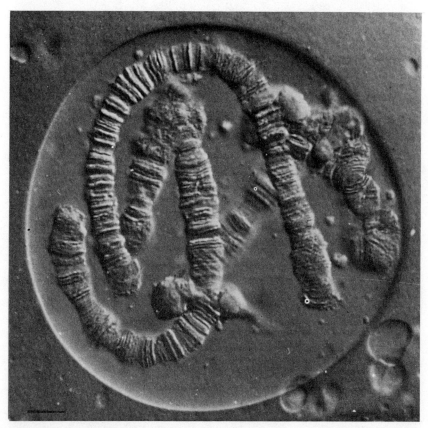

Figure 7-2. Polytene nucleus of *Chironomus tentans* as seen in light microscope using Nomarski optics. Bar represents 10 μm. *(From Macgregor, H., and Varley, J. V., 1983, p. 106; copyright © 1983 by John Wiley & Sons Ltd.; reprinted by permission.)*

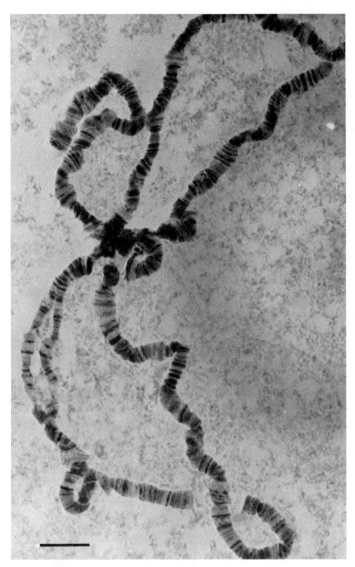

Figure 7-3. Polytene chromosomes of *Drosophila melanogaster* salivary glands. Note the banding pattern on chromosome arms. The individual chromosomes aggregate at a define place to form a chromocenter. Bar represents 10 μm. *(From Lewin, 1980, p. 461; copyright © 1980 by John Wiley & Sons, Inc.; photograph by J. J. Bonner reprinted by permission.)*

tionized research in the field of polytene chromosomes. The first technique, generally known as in situ hybridization, was introduced by Gall and Pardue (1969), followed by the similar procedures of John et al. (1969) and Buongiorno-Nardelli and Amaldi (1970). These authors demonstrated that it is possible to hybridize labeled DNA or RNA probes to cytological preparations and subsequently to detect sites of hybridization by autoradiography. Since each polytene chromosome consists of a few hundred to a thousand parallel double stranded DNA molecules and has excellent intrachromosomal markers,

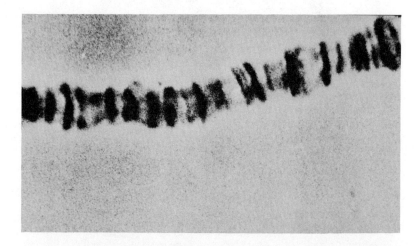

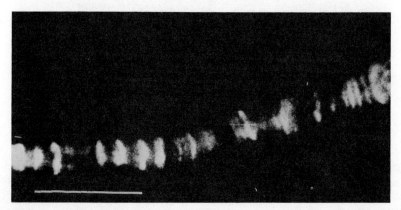

Figure 7-4. Visualization of histone H1 distribution in polytene chromosomes of *Drosophila melanogaster*. Upper panel (*4a*) shows the phase contrast picture of the chromosome; the lower panel (*4b*) shows the pattern of immunofluorescence after staining with antibodies against histone H1. (From Jamrich et al., 1977)

the bands, no other chromosomes were more suited for the analysis of DNA in this fashion. Initially, this analysis was hampered by the difficulties in obtaining purified segments of DNA for use as probes, but with the aid of the newly developed cloning techniques a powerful tool was created. Virtually any DNA sequence can be cloned and its precise chromosomal location determined. Since the introduction of this method a large number of genes, as well as highly and moderately repeated sequences, have been localized. The replacement of radioactive probes by biotin labeled probes which bind rabbit anti-biotin antibodies, and can be detected either fluorimetrically, by using fluorescein-labeled anti-rabbit IgG antibodies, or cytochemically, by using an anti-rabbit IgG antibody conjugated to horseradish peroxidase, enables precise localization of a given DNA sequence in a matter of hours (Langer-Safer et al., 1982).

The second technique which deserves mention is the localization of proteins on polytene chromosomes using antibodies. This method, indirect immunofluorescence, takes advantage of the fact that the antibodies frequently recognize in cytological preparations the antigens they were raised against. Their binding sites are identified by the reaction of second antibod-

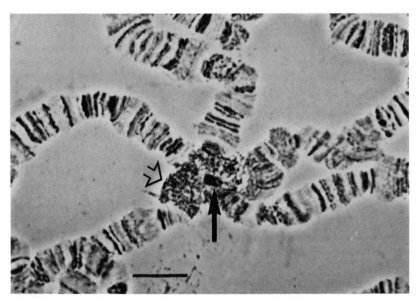

Figure 7-5. A large magnification of a chromocenter of *Drosophila virilis*. The α-heterochromatin is labeled with a solid arrow, the B-heterochromatin with an open arrow. Bar represents 10 μm *(From Gall et al., 1971, p. 321; copyright © 1971 by Springer Verlag; reprinted by permission.)*

ies directed against the first. The second antibodies are usually labeled with a fluorescent or otherwise traceable compound, and therefore can be detected in the light or electron microscope. Indirect immunofluorescence, which was first used on polytene chromosomes by Desai et al. (1972), and which was reintroduced by Silver and Elgin (1976) and Alfageme et al. (1976), offers the advantage of correlating protein distribution with chromosomal regions of defined function. The technique, though very useful, is limited by the fact that it is tedious to purify individual proteins and, in most cases, virtually impossible to deduce their function.

Finally, the third important technical breakthrough was the development of the *Drosophila* transformation system by Spradling and Rubin (1982). This relatively recent method will without any doubt have significant influence on the study of polytene chromosomes. The possibility of the introduction of novel sequences or the reintroduction of modified *Drosophila* sequences into the genome will allow detailed studies on the structure of polytene chromosomes in transformed individuals. It should be possible to determine the influence of exogeneously introduced sequences on the formation of morphologically distinct regions (i.e., bands, interbands, etc.).

MOLECULAR BASIS OF THE BANDING PATTERN

The fact that the polytene chromosomes are composed of hundreds of DNA molecules aligned side by side, of which at least some seem to extend the entire length of the chromosome (Beermann and Pelling, 1965; Kavenoff and Zimm, 1973), could lead one to the conclusion that the amount of DNA in different chromosomal regions is the same. This is, however, not the case. Beermann (1972) concluded that up to 95% of the DNA is present in the bands. This is probably an overestimate, and the actual percentage is likely to be closer to 75% (Laird, 1980); nevertheless, most of the DNA is located in the bands. The reason for the discrepancy in the DNA content between bands and interbands was puzzling for many years, and two models were proposed to explain this phenomenon. The first was based on the assumption that the bands are replicated to a larger degree than the neighboring interbands (Sorsa, 1974; Laird, 1980; Laird et al., 1980). The second model suggested a higher degree of folding of the basic fiber in the bands than in the interbands (DuPraw and Rae, 1966; and earlier workers in general). The hypothesis that the banding pattern is a consequence of a differential replication of bands and interbands has been critically tested recently. Spierer and Spierer (1984), and Spierer et al. (1983) analyzed a 315,000 base pair long segment of *Drosophila* DNA which encompasses 13 bands and their adjacent interbands. They found by Southern blot analysis (Southern, 1975) that the different restriction fragments, 84 in all, have a very similar degree of polyteny. Similarly, Lifshytz (1983) demonstrated that randomly selected unique sequences which map to different positions on the polytene chromosomes are replicated to the same degree.

However, not all of the DNA sequences in polytene chromosomes have the same degree of polyteny. Satellite sequences, ribosomal RNA genes, histone genes, bithorax complex sequences, as well as the entire male Y-chromosome are underreplicated in *Drosophila* polytene tissues, but this underreplication does not seem to be related to the proposed role in formation of bands and interbands. For example, the highly repetitive satellite sequences, which have been shown by Gall et al. (1971) to be underreplicated, are located in the transcriptionally inactive a-heterochromatin of the chromocenter. The ribosomal genes that are also underreplicated in many species, though generally to a lesser extent than the satellite type repetitive sequences (Henning and Meer, 1971; Spear and Gall, 1973; for review see Brutlag, 1980), are located in the heterochromatic chromocenter as well. The histone gene family of *Drosophila melanogaster* which is a tandem array of 100–300 genes (Lifton et al., 1978, Chernyshev et al., 1980) is located in the chromosomal region 39DE of Bridges map (Pardue et al., 1977) and is also underreplicated (Lifshytz, 1983), as are the bithorax complex genes (Spierer et al., as cited in Spierer and Spierer, 1984). Cytological analysis of these two gene complexes shows that although they are not located in the heterochromatic chromocenter, they are either close to or span a chromosomal constriction. Chromosomal constrictions are regions of intercalary heterochromatin (see Beermann, 1962), which might contain sequences responsible for the underreplication of the neighboring DNA. In summary, the underreplication of all the aforementioned sequences is either due to the fact that they are heterochromatic, that is, do not proportionally replicate, or that they are in the proximity of such sequences.

Overreplication of DNA sequences prior to their transcription has been reported in *Sciara* and *Rhynchosciara* (Breuer and Pavan, 1955; Crouse and Keyl, 1968), though this phenomenon seems to be limited to *Sciaridae*. The reasons for this overreplication are not understood, but it is extremely unlikely that this overreplication, as well as the previously described underreplication, has anything to do with generation of band-interband morphology. Consequently, it appears likely that the formation of bands and interbands is due to a differential folding of the basic chromosomal fiber. This is supported by evidence from the electron microscopic examination of the chromatin from polytene chromosomes, which shows that the basic chromatin fiber has a "beads-on-a-string" appearance and is 10 nm in diameter. This fiber is a result of the association of DNA with basic nuclear proteins—the histones (The molecular basis of this association is described in detail in chap. 1 of this volume). In vivo, the 10 nm fiber is coiled to a higher order structure, which is known as the thick chromatin fiber or chromatid, and which is 25 nm in diameter. This fiber is present in an extended form (25 nm in diameter) in the interbands, but is highly coiled in the bands (Sass, 1980). This differential coiling of the 25 nm fiber is most likely the basis for the banded appearance of polytene chromosomes. Gentle uncoiling of the condensed chromatin of the bands reveals that in the band region each chromatid forms a single loop which is condensed to a compact structure. The size of the loop appears to be proportional to the size of the given band. It is uncertain whether the distribution of the condensed chromatin within the band is uniform; it has been suggested that this condensed chromatin forms a toroidal structure (Mortin and Sedat, 1982).

BANDS AND GENES

The banding pattern is one of the most intriguing features of polytene chromosomes. It is remarkably constant within a particular species, but shows striking differences among species. This reproducibility of banding pattern led the early investigators to the conclusion that the more condensed regions of the chromosomes, namely the bands, correspond to basic genetic functional units—the genes (Bridges, 1935; Painter, 1934; Muller and Prokofyeva, 1935). This notion was supported by the fact that *Drosophila* has approximately the same number of observable bands as the presumed number of genes (5,000). Genetic analysis of well defined regions revealed that the number of complementation groups in the subsegments of chromosomes correlates with the number of bands in the given region. Judd et al. (1972), for example, found 14 complementation groups in the region 3A2-3C2, which spans 16 bands. Lefevre (1974), Woodruff and Ashburner (1979), Gausz et al. (1979, 1981), and Hilliker et al. (1980b), analysing different regions of chromosomes, came to similar conclusions. In recent years, however, evidence emerged suggesting rather strongly that the one band-one gene hypothesis might not have general validity. First of all, there is good genetic evidence that certain bands contain more than one complementation group (Young and Judd, 1978; Wright et al., 1981; Zhimulev et al., 1981). Second, Hall et al. (1983) found, using Northern blot analysis, that one of the bands in the 87 E region contains four transcriptional units. Third, the 165 5S genes (Procunier and Tartof, 1975; Procunier and Dunn, 1978; Wimber and Steffensen, 1970), as well as the 100-300 histone genes (Lifton et al., 1978; Chernyshev et al., 1980), certainly span many fewer bands than would be required if there was a one-to-one correlation between the genes and bands.

Taking all these arguments at face value, it is hard to avoid the impression that the one band-one gene hypothesis is at least an oversimplification, if not a complete misconception. It appears more likely that bands are structural domains which contain one or few genes (transcriptional units) much in the way loops on lampbrush chromosomes do (for review of lampbrush chromosomes see chap. 5, this volume). The pattern of domains (bands or loops) might represent the most energetically favorable packaging of the given chromosomal DNA. This packaging naturally does have an influence on transcription and replication units, but strictly speaking does not represent either of them. Indeed, a band itself might not be a functional unit of any kind.

TRANSCRIPTION ON POLYTENE CHROMOSOMES

Although polytene chromosomes display constant banding patterns, they are by no means static structures. At certain physiological stages, the bands swell and form diffuse structures known as puffs (Poulson and Metz, 1938). Beermann (1952) demonstrated that puffs are generated by local uncoiling of the chromatin of the bands. The highly condensed chromatin of the bands unravels first into the 25 nm chromatin fiber and then into the 10 nm fiber. During the most active transcription the 10 nm fiber loses its beaded appearance, and is present as a 5 nm smooth fiber. After the transcription ceases, a condensation takes place in the reversed order (Andersson et al., 1982; 1984).

Puffs can be generated from single or multiple bands. They are the loci of most active RNA synthesis (Pelling, 1959; Sirlin, 1960), exhibit high levels of RNA polymerase (Plagens et al., 1976; Jamrich et al., 1977; Greenleaf et al., 1978), and their size is proportional to the intensity of RNA transcription (Pelling, 1964; Edstrom and Daneholt, 1967). In *Chironomus*, the largest puffs, called Balbiani rings, exhibit transcriptional levels two or three orders of magnitude higher than other puffs, and can have extraordinary proportions (Fig. 7-6). Puffs on polytene chromosomes vary from tissue to tissue (Beermann, 1952, 1961; Mechelke, 1953, 1958, 1963; Becker, 1959; Berendes, 1965a, 1965b, 1967) and the puffing pattern is developmentally regulated (Clever and Karlson, 1960; Panitz, 1960, 1964; Clever, 1961, Ashburner, 1967). Causal correlation between appearance of certain puffs and production of certain developmentally regulated proteins has been made (Baudish and Panitz, 1968; Grossbach, 1969, 1973; Korge, 1975, 1977; Akam et al., 1978). Some of the puffs can be induced by environmental changes, such as elevated temperature (Ritossa, 1962). The treatment of larvae with heat results in the induction of so-called heat shock puffs, redistribution of RNA polymerase on the chromosomes (Jamrich et al., 1977; Greenleaf et al., 1978), the massive production of heat shock RNAs, and a reduction in transcription at nonheat shock loci (Ritossa, 1964; Berendes, 1968; Tissieres

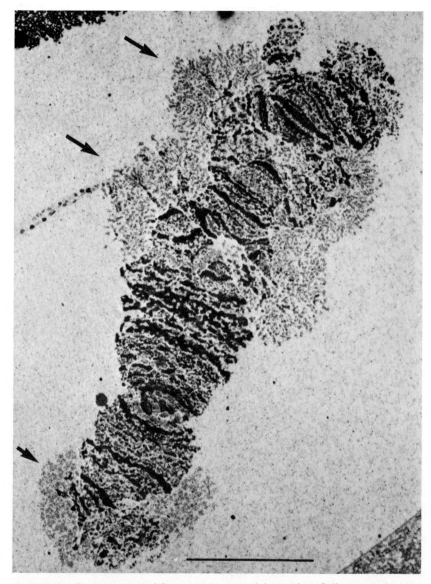

Figure 7-6. Chromosome 4 of *Chironomus tentans* exhibiting three Balbiani rings (arrows). Bar represents 10 μm. *(From Daneholt, 1975, p. 1; copyright © 1975 by The M.I.T. Press.)*

et al., 1974; for review see Ashburner and Bonner, 1979). Puffing is not limited to *Drosophila* sequences. Sequences foreign to *Drosophila* will puff in salivary gland cells of transformed larvae if introduced into the genome preceded by a strong promoter (Lis et al., 1983). Nevertheless, not all transcription results in the formation of a puff. 5S and histone genes are known to be transcribed—their location is well established—but these loci do not puff.

What is the difference between the transcription of genes which generate a puff and those which do not? The definitive answer is not known at present, but it is tempting to speculate that the degree of puffing is not only related to the intensity of transcription but also depends on the size of the transcribed DNA. The Balbiani rings, for example, generate an enormous 75S RNA which is more reminiscent of the giant transcripts found on amphibian lampbrush chromosome loops than what is generally considered to be the size of an average somatic transcriptional unit. The similarity to lampbrush chromosomes does not end here. The transcription of 5S genes on lampbrush chromosomes does not normally result in the formation of a loop either, presumably because the transcription unit is too small. The larger the transcription unit, the larger the loop. The same principle might be involved in the puffing process. Alternatively, a large puff might be generated if a few neighboring transcription units are activated together, forming a transcriptionally active region long enough to be microscopically visible. In support of this alternative, it has been demonstrated that more than one gene may be transcribed in a puff (Crowley et al., 1983). This principle is also involved in the formation of lampbrush loops. Considering that the transcription of 5S and histone genes does not result in the formation of a puff, one might rightfully ask what percentage of genes are actually transcribed in the puffed configuration.

A number of authors in the past two decades have demonstrated that the puffs are not the only sites of active transcription (Zhimulev and Belyaeva, 1975; Semeshin et al., 1979; and others). The additional transcriptional sites were identified as the interbands, which were proposed to be either regulatory sequences or sites of low level transcription of genes involved in housekeeping functions (Crick, 1971; Paul, 1972). Further support for the hypothesis that the interbands are transcriptionally active came from the work of Jamrich et al. (1977), who demonstrated that antibodies against RNA polymerase II bind not only to obvious puffs but also to other regions of chromosomes, which were identified as interbands. Although their technique was not refined enough to distinguish every single interband, they correctly recognized the positive correlation between the immunofluorescence pattern and decondensed regions of chromosomes. Their observations were confirmed at the light and electronmicroscopic level using either monoclonal antibodies against the same enzyme (Kramer et al., 1980; Sass and Bautz, 1982a) or in conjunction with ³H uridine incorporation (Sass, 1982; Sass and Bautz, 1982b). Further supporting evidence came from work of Skaer (1977), who presented data to suggest that the interbands contain ribonucleoproteins (RNPs). Saumweber et al. (1980) and Christensen et al. (1981), demonstrated that at least some of the RNA binding proteins are located in the interband regions of polytene chromosomes.

All this evidence is rather suggestive, but it is of course possible to argue that most of the light microscopically visible interbands are actually tiny puffs and then the argument becomes semantic. Nevertheless, it appears safe to say that not all transcription is taking place in puffs visible in the light microscope. Now it is possible to turn the question around and ask whether puffing is always accompanied by transcription. Most of the published evidence certainly suggests so (Pelling, 1964; Berendes, 1968; Zhimulev and Belyaeva, 1975). Induction of puffs can be prevented by inhibitors of transcription (Beerman, 1964, 1971). Although termination of transcription and the disappearance of RNA polymerase from the puff does not result in immediate regression of that puff (Ashburner, 1972; Jamrich et al., 1977), generation of a puff without transcription is questioned. More recent evidence from E. Meyerowitz's laboratory (pers. commun.) suggests that a puff can be formed without actual transcription.

DISTRIBUTION OF PROTEINS ON POLYTENE CHROMOSOMES

DNA in polytene chromosomes is complexed with the histones and a large number of nonhistone proteins. The association with histones results in a repeated "nucleosomal" structure of the nucleoprotein fiber (Hewish and Burgoyne 1973; Kornberg, 1974; Kornberg and Thomas, 1974; Noll, 1974; Oudet et al., 1975; for review see chap. 1 of this volume). Two molecules each of histone H2a, H2b, H3 and H4 are complexed with about 200 base pairs of DNA to form a nucleosome. They are separated by a variable stretch of DNA associated with one or more histone H1 molecules. The nucleosomal arrangement of the DNA can be visualized either by electron microscopy where the nucleosomes appear as a beaded structure (Olins and Olins, 1974), or indirectly by the analysis of micrococcal nuclease resistant fragments in a digest of chromatin separated by agarose or polyacrylamide gel eletrophoresis (Noll, 1974). Nucleosomes are evenly distributed along the entire length of the DNA, a conclusion based on the observation that the staining of polytene chromosomes for histones result in the same banding pattern as the staining for DNA (Swift, 1964), and immunofluorescence studies using antibodies against histone proteins (Desai et al., 1972; Alfageme et al., 1976; Jamrich et al., 1977; Silver and Elgin, 1977; Kurth et al., 1978). Figure 7-4 shows polytene chromosomes stained with antibodies against histone H1. The comparison of the fluorescence pattern (4b) with the phase contrast picture (4a) shows that the amount of histone H1 is proportional to the amount of material in a given band or interband. The same can be concluded for the other four histones. The distribution of histones is constant, and, in contrast to nonhistone proteins, little if any redistribution can be found during transcription of chromosomal loci (Silver and Elgin, 1976; Alfageme et al., 1976; Jamrich et al., 1977).

The distribution of nonhistone proteins has been studied using polyclonal and monoclonal antibodies. A number of nonhistone proteins follow closely the redistribution of transcriptional loci. These proteins can be divided into two groups. The first group is comprised of proteins postulated to be directly involved in the process of transcription. To this class belong the already mentioned RNA polymerase, topoisomerase I (Steiner et al., 1984), and other nonhistone proteins of unknown function, which were either isolated from DNase I digests of active chromatin (Mayfield et al., 1978) or deduced to be involved in transcription from antibody staining patterns (Silver and Elgin, 1977, 1978; Howard et al., 1981). Antibodies against the latter class of proteins usually stain a variable number of chromosomal sites during or prior to puffing.

The second group of proteins which react preferentially with transcriptionally active regions are the RNA binding proteins. These proteins are either involved in RNA packaging or in RNA processing. Antibodies against these proteins react either with most (if not all) transcriptional sites (Christensen et al., 1981, Kabisch and Bautz, 1983), or decorate only a relatively small subset of active loci (Dangli et al., 1983; Risau et al., 1983). There are also proteins found in nonrandom association with chromosomes which cannot be positively correlated with active transcription. For example, antibodies against HMG-1 (high mobility group 1) proteins stain primarily the chromosomal bands (Kurth and Bustin, 1981), and the antibody described by Will and Bautz (1980) recognizes proteins that preferentially bind to the centromeric heterochromatin of *Drosophila*.

The use of indirect immunofluorescence is not limited to localization of proteins. Any substance which is antigenic can be localized even though the interpretation is not always straightforward. For example, attempts have been made to identify the regions of Z-DNA in polytene chromosomes (Nordheim et al., 1981; Lemeunier et al., 1982; Arndt-Jovin et al., 1983). Z-DNA can assume a left-handed helical conformation (Wang et al., 1981; Drew and Dickerson, 1981) and therefore might play an important role in gene activation or inactivation. Unfortunately, the structure of DNA in polytene chromosomes seems to be very sensitive to fixation conditions (Hill and Stollar, 1983), and consequently data generated in this field as of yet have not yet provided definitive information about the distribution of Z-DNA on polytene chromosomes in vivo.

This and many other questions about polytene chromosomes remain unanswered despite significant progress in the last few decades. However, the revived interest in polytene chromosomes offers the hope that we will find out the answers in the not too distant future.

I would like to thank Drs. I. B. Dawid, J. G. Gall, K. A. Mahon, and P. M. M. Rae for critical reading of the manuscript and valuable suggestions, as well as Drs. J. J. Bonner, B. Daneholt, J. G. Gall, and H. Macgregor for original photographs.

REFERENCES

Agard, D. A.; and Sedat, J. W. (1983). Three-dimensional architecture of a polytene nucleus. *Nature* **302**, 676-681.

Akam, M. E.; Roberts, D. B.; Richards, G. P.; and Ashburner, M. (1978). Drosophila: the genetics of two major larval proteins. *Cell* **13**, 215-226.

Alfageme, C. R.; Rudkin, G. T.; and Cohen, L. H. (1976). Locations of chromosomal proteins in polytene chromosomes. *Proc. Natl. Acad. Sci. (U.S.A.)* **73**, 2038-2042.

Andersson, K.; Mahr, R.; Bjorkroth, B.; and Daneholt, B. (1982). Rapid reformation of thick chromosome fiber upon completion of RNA synthesis at the Balbiani ring genes in *Chironomus tentans*. *Chromosoma* **87**, 33-48.

Andersson, K.; Bjorkroth, B.; and Daneholt, B. (1984). Packing of a specific gene into higher order structures following repression of RNA synthesis. *J. Cell Biol.* **98**, 1296-1303.

Arndt-Jovin, D. J.; Robert-Nicoud, M.; Zarling, D. A.; Greider, C.; Weimer, E.; and Jovin, T. M. (1983). Left-handed Z-DNA in bands of acid-fixed polytene chromosomes. *Proc. Natl. Acad. Sci. (U.S.A.)* **80**, 4344-4348.

Ashburner, M. (1967). Autosomal puffing patterns in a laboratory stock of *Drosophila melanogaster*. *Chromosoma* **21**, 398-428.

Ashburner, M. (1972). Ecdysone induction of puffing in polytene chromosomes of *D. melanogaster*. Effects of inhibitors of RNA synthesis. *Exp. Cell. Res.* **71**, 433-440.

Ashburner, M.; and Bonner, J. J. (1979). The induction of gene activity in *Drosophila* by heat shock. *Cell* **17**, 241-254.

Balbiani, E. G. (1881). Sur la structure du noyau des cellules salivaires chez les larves de *Chironomus*. *Zool. Anz.* **4**, 637-641.

Baudisch, W.; and Panitz, R. (1968). Kontrolle eines biochemischen Merkmals in den Speicheldrusen von *Acricotopus lucidus* durch einen Balbiani ring. *Exp. Cell. Res.* **49**, 470-476.

Bauer, H. (1935). Der Aufbau der Chromosomen aus den Speicheldrusen von *Chironomus Thummi* Kiefer (Untersuchungen an den Riesenchromosomen der Dipteren I). *Z. Zellforsch.* **23**, 280-313.

Becker, H. J. (1959). Die Puffs der Speicheldrusenchromosomen von *Drosophila melanogaster*. 1. Mitteilung. Beobachtungen zum Verhalten des Puffmusters in Normalstamm und bei zwei Mutanten, Giant und Lethal-Giant-Larvae. *Chromosoma* **10**, 654-678.

Beermann, W. (1952). Chromosomenkonstanz und spezifische Modifikationen der Chromosomenstruktur in der Entwicklung von *Chironomus tentans*. *Chromosoma* **5**, 139-198.

Beermann, W. (1961). Ein Balbiani-Ring als Locus einer Speicheldrusemutation. *Chromosoma* **12**, 1-25.

Beermann, W. (1962). *Riesenchromosomen. Protoplasmatologia, Handbuch der Protoplasmaforschung.* Band VI, D. Springer-Verlag, Vienna.
Beerman, W. (1964). Control of differentiation at the chromosomal level. *J. Exp. Zool.* **157,** 49-62.
Beermann, W. (1971). Effect of α-amanitin on puffing and intranuclear RNA synthesis in *Chironomus* salivary glands. *Chromosoma* **34,** 152-167.
Beerman, W. (1972). In Results and Problems in Cell Differentiation. (Beerman, W.; Reinert, J.; and Ursprung, H.; eds.) vol. 4, Springer-Verlag, Berlin.
Beermann, W.; and Pelling, C. (1965). 3H-Thymidin Markierung einzelner Chromatiden in Riesenchromosomen. *Chromosoma* **16,** 1-21.
Berendes, H. D. (1965a). Salivary gland functions and chromosomal puffing patterns in *D. hydei. Chromosoma* **17,** 35-77.
Berendes, H. D. (1965b). The induction of changes in chromosomal activity in different polytene types of cell in *D. hydei. Dev. Biol.* **11,** 371-384.
Berendes, H. D. (1967). The hormone ecdysone as effector of specific changes in the pattern of gene activities of *Drosophila hydei. Chromosoma* **29,** 118-130.
Berendes, H. D. (1968). Factors involved in the expression of gene activity in polytene chromosomes. *Chromosoma* **24,** 418-437.
Berendes, H. D.; and Keyl, H. G. (1967). Distribution of DNA in heterochromatin and euchromatin of polytene nuclei of *D. hydei. Genetics* **57,** 1-13.
Breuer, M. E.; and Pavan, C. (1955). Behavior of polytene chromosomes of *Rhynchosciara angelae* at different stages of larval development. *Chromosoma* **7,** 371-386.
Bridges, C. B. (1935). Salivary chromosome map with a key to the banding of the chromosomes of *D. melanogaster. J. Hered.* **26,** 60-64.
Brutlag, D. (1980). Molecular arrangement and evolution of heterochromatic DNA. *Annu. Rev. Genet.* **14,** 121-144.
Buongiorno-Nardelli, M.; and Amaldi, F. (1970). Autographic detection of molecular hybrids between rRNA and DNA in tissue sections. *Nature* **225,** 946-948.
Caspersson, T. (1940). Die Eiweissverteilung in den Strukturen des Zellkerns. *Chromosoma* **1,** 562-604.
Chernyshev, A. I.; Bashkirov, V. N.; Leibovitch, B. A.; and Khesin, R. B. (1980). Increase in the number of histone genes in case of their deficiency in *Drosophila melanogaster. Mol. Gen. Genet.* **178,** 663-668.
Christensen, M. E.; LeStourgeon, M.; Jamrich, M.; Howard, G. C.; Serunian, L. A.; Silver, L. M.; and Elgin, S. C. (1981). Distribution studies on polytene chromosomes using antibodies directed against hnRNP. *J. Cell Biol.* **90,** 18-24.
Clever, U. (1961). Genaktivitaten in den Riesenchromosomen von *Chironomus tentans* und ihre Beziehung zur Entwicklung. I. Genaktivierungen durch Ecdyson. *Chromosoma* **12,** 607-675.
Clever, U.; and Karlson, P. (1960). Induktion von Puff-Veranderungen in den Speicheldrusenchromosomen von *C. tentants* durch Ecdysone. *Exp. Cell Res.* **20,** 623-626.
Crick, F. (1971). General model for the chromosomes of higher organisms. *Nature* **234,** 25-27.
Crouse, H. V.; and Keyl, H. G. (1968). Extra replication in the "DNA puffs" of *Sciara coprophila. Chromosoma* **25,** 357-364.
Crowley, T. E.; Bond, M. W.; and Meyerowitz, E. M. (1983). The structural genes for three *Drosophila* glue proteins reside at a single polytene chromosome puff locus. *Mol. Cell Biol.* **3,** 623-634.
Danehold, B. (1975). Transcription in Polytene Chromosomes. *Cell* **4,** 1-9.
Dangli, A.; Grond, C.; Kloetzel, P.; and Bautz, E. K. F. (1983). Heat-shock puff 93D from *Drosophila melanogaster*: accumulation of a RNP-specific antigen associated with giant particles of possible storage function. *EMBO J.* **2,** 1747-1751.
Desai, L. S.; Pothier, L.; Foley, G. E.; and Adams, R. A. (1972). Immunofluorescent labeling of chromosomes with antisera to histones and histone fractiones. *Exp. Cell Res.* **70,** 468-471.

Drew, H. R., and Dickerson, R. E. (1981). Conformation and dynamics in a Z-DNA tetramer. *J. Mol. Biol.* **152**, 723-736.

DuPraw, E. J.; and Rae, P. M. M. (1966). Polytene chromosome structure in relation to the "folded fibre" concept. *Nature* **212**, 598-600.

Edstrom, J.; and Daneholt, B. (1967). Sedimentation properties of the newly synthesized RNA from isolated nuclear components of *Chironomus tentans* salivary gland cells. *Exp. Cell Res.* **57**, 205-210.

Gall, J. G.; and Pardue, M. L. (1969). Formation and detection of RNA-DNA hybrid molecules in cytological preparations. *Proc. Natl. Acad. Sci. (U.S.A.)* **63**, 378-383.

Gall, J. G.; Cohen, E. H.; and Polan, M. L. (1971). Repetitive DNA sequences in *Drosophila*. *Chromosoma* **33**, 319-344.

Gausz, J.; Bencze, G.; Gyurkovics, H.; Ashburner, M.; Ish-Horowitz, D.; and Holden, J. J. (1979). Genetic characterization of the 87C region of the third chromosome of *Drosophila melanogaster*. *Genetics* **93**, 917-934.

Gausz, J.; Gyurkovics, H.; Bencze, G.; Awad, A. A. M.; Holden, J. J.; and Ish-Horowitz, D. (1981). Genetic characterization of the region between 86F1,2 and 87B15 on chromosome 3 of *Drosophila melanogaster*. *Genetics* **98**, 775-789.

Greenleaf, A. L.; Plagens, U.; Jamrich, M.; and Bautz, E. K. F. (1978). RNA polymerase B (or II) in heat induced puffs of *Drosophila* polytene chromosomes. *Chromosoma* **65**, 127-136.

Grossbach, U. (1969). Chromosomen-Aktivitat und biochemische Zelldifferenzierung in den Speicheldrusen von *Camptochironomus*. *Chromosoma* **28**, 136-187.

Grossbach, U. (1973). Chromosome puffs and gene expression in polytene cells. *Cold Spring Harbor Symp. Quant. Biol.* **38**, 619-627.

Hall, L. M.; Mason, P. J.; and Spierer, P. (1983). Transcripts, genes and bands in 315,000 base-pairs of *Drosophila* DNA. *J. Mol. Biol.* **169**, 83-96.

Heitz, E. (1934). Uber α-und β-Heterochronmatin sowie Konstanz und Bau der Chromomeren bei *Drosophila*. *Biol. Zb.* **54**, 588-609.

Heitz, E.; and Bauer, H. (1933). Beweise fur die Chromosomennatur der Kernschleifen in den Knauelkernen von *Bibio hortulanus* L. *Z. Zellforsch.* **17**, 67-82.

Henning, W.; and Meer, B. (1971). Reduced polyteny of rRNA cistrons in giant chromosomes of *D. hydei*. *Nature New Biol.* **23**, 70-72.

Hewish, D. R.; and Burgoyne, L. A. (1973). Chromatin substructure. The digestion of chromatin DNA at regularly spaced sites by a nuclear DNase. *Biochem. Biophys. Res. Commun.* **52**, 504-510.

Hill, R. J.; and Stollar, B. D. (1983). Dependence of Z-DNA antibody binding to polytene chromosomes on acid fixation and DNA torsional strain. *Nature* **305**, 338-340.

Hilliker, A. J.; Appels, R.; and Schalet, A. (1980a). The genetic analysis of *D. melanogaster* heterochromatin. *Cell* **21**, 607-619.

Hilliker, A. J.; Clark, S. H.; Chovnick, A.; and Gelbart, W. M. (1980b). Cytogenetic analysis of the chromosomal region immediately adjacent to the *rosy* locus in *Drosophila melanogaster*. *Genetics* **95**, 95-110.

Hinton, T. (1945). A study of chromosome ends in salivary gland nuclei of *Drosophila*. *Biol. Bull.* **88**, 144-165.

Howard, G. C.; Abmayr, S. M.; Shinefeld, L. A.; Sato, V. L.; and Elgin, S. C. R. (1981). Monoclonal antibodies against a specific nonhistone chromosomal protein associated with active genes. *J. Cell Biol.* **88**, 219-225.

Hughes-Schrader, S. (1943). Polarization, kinetochore movements, and bivalent structure in the meiosis of male mantids. *Biol. Bull.* **85**, 265-300.

Jamrich, M.; Greenleaf, A. L.; and Bautz, E. K. F. (1977). Localization of RNA polymerase in polytene chromosomes of *Drosophila*. *Proc. Natl. Acad. Sci. (U.S.A.)* **74**, 2079-2083.

John, H.; Birnstiel, M.; and Jones, K. (1969). RNA-DNA hybrids at the cytological level. *Nature* **223**, 582-587.

Judd, B. H.; Shen, M. W.; and Kaufmann, T. C. (1972). The anatomy and function of a segment of the X chromosome of *Drosophila melanogaster. Genetics* **71**, 139-156.

Kabisch, R.; and Bautz, E. K. F. (1983). Differential distribution of RNA polymerase B and nonhistone proteins in polytene chromosomes of *Drosophila melanogaster. EMBO J.* **2**, 395-402.

Kavenoff, R.; and Zimm, B. H. (1973). Chromosome sized DNA molecules from *Drosophila. Chromosoma* **41**, 1-28.

King, R. L.; and Beams, H. W. (1934). Somatic synapsis in Chironomus with special reference to the individuality of the chromosomes. *J. Morphol.* **56**, 577-586.

Korge, G. (1975). Chromosome puff activity and protein synthesis in larval salivary glands of *D. melanogaster. Proc. Natl. Acad. Sci. (U.S.A.)* **72**, 4550-4554.

Korge, G. (1977). Direct correlation between a chromosome puff and the synthesis of a larval salivary protein in *Drosophila melanogaster. Chromosoma* **62**, 155-174.

Kornberg, R. D. (1974). Chromatin structure: a repeating unit of histones and DNA. *Science* **184**, 868-871.

Kornberg, R. D.; and Thomas, J. O. (1974). Chromatin structure: oligomers of histones. *Science* **184**, 865-868.

Kramer, A.; Haars, R.; Kabish, R.; Will, H.; Bautz, F. A.; and Bautz, E. K. F. (1980). Monoclonal antibody directed against RNA polymerase II of *Drosophila melanogaster. Mol. Gen. Genet.* **180**, 193-199.

Kurth, P. D.; and Bustin, M. (1981). Localization of chromosomal protein HMG-1 in polytene chromosomes of *Chironomus thummi. J. Cell Biol.* **89**, 70-77.

Kurth, P. D.; Moudrianakis, E. N.; and Bustin, M. (1978). Histone localization in polytene chromosomes by immunofluorescence. *J. Cell Biol.* **78**, 910-918.

Laird, C. D. (1980). Structural pardox of polytene chromosomes. *Cell* **22**, 869-874.

Laird, C. D.; Wilkinson, L.; Johnson, D.; and Sandstrom, C. (1980). In *Chromosomes Today*, (M. Bennett, M. Bobrow, and G. Hewitt, eds.), vol. 7, pp. 74-83, George, Allen, and Unwin, London.

Lakhotia, S. C. (1974). EM autoradiographic studies on polytene nuclei of *D melanogaster*. III. Localisation of nonreplicating chromatin in the chromocentre heterchromatin. *Chromosoma* **46**, 145-160.

Lakhotia, S. C.; and Jacob, J. (1974). EM autoradiographic studies on polytene nuclei of *D. melanogaster*. II. Organization and transcriptive activity of the chromocentre. *Exp. Cell Res.* **86**, 253-263.

Langer-Safer, P. R.; Levine, M.; and Ward, D. C. (1982). Immunological method for mapping genes on *Drosophila* polytene chromosomes. *Proc. Natl. Acad. Sci. (U.S.A.)* **79**, 4381-4385.

Lefevre, G., Jr. (1974). The relationship between genes and polytene chromosome bands. *Annu. Rev. Genet.* **8**, 51-62.

Lemeunier, F.; Derbin, C.; Malfoy, B.; Leng, M.; and Taillandier, E. (1982). Identification of left-handed Z-DNA by indirect immunofluorescence in polytene chromosomes of *Chironomus thummi thummi. Exp. Cell Res.* **141**, 508-513.

Lewin, B. (1980). *Gene Expression*, vol. 2, 2nd ed., Wiley, New York.

Lifschytz, E. (1983). Sequence Replication and banding organization in the polytene chromosomes of *Drosophila melanogaster. J. Mol. Biol.* **164**, 17-34.

Lifton, R. P.; Goldberg, M. L.; Karp, R. W.; and Hogness, D. S. (1978). *Cold Spring Harbor Symp. Quant. Biol.* **42**, 1047-1051.

Lis, J. T.; Simon, J. A.; and Sutton, C. A. (1983). New heat shock puffs and beta-galactosidase activity resulting from transformation of *Drosophila* with an hsp70-lacZ hybrid gene. *Cell* **35**, 403-410.

Macgregor, H.; and Varley, J. W. (1983). *Working with animal chromosomes.* Wiley Ltd., Chichester, England.

Mathog, D.; Hochstrasser, M.; Gruenbaum, Y.; Saumweber, H.; and Sedat, J. (1984). Characteristic folding pattern of polytene chromosomes in *Drosophila* salivary gland nuclei. *Nature* **308**, 414-421.

Mayfield, J. E.; Serunian, L. A.; Silver, L. M.; and Elgin, S. C. R. (1978). A protein released by DNase I digestion of *Drosophila* nuclei is preferentially associated with puffs. *Cell* **14**, 539-544.

Mechelke, F. (1953). Reversible Strukurmodifikationen der Speicheldrusenchromosomen von *Acricotopus lucidus*. *Chromosoma* **5**, 511-543.

Mechelke, F. (1958). *The timetable of physiological activity of several loci in the salivary gland chromosomes of* Acricotopus lucidus. Proceedings of the 10th International Congress on Genetics, vol. II., University of Toronto Press, Toronto.

Mechelke, F. (1963). *Spezielle Funktionszustande des genetischen Materials.* Wissenschaftliche Konfer enz der Gesselschaft der Deutschen Naturforscher und Artzte, Springer-Verlag, Berlin, Gottingen-Heidelberg, pp. 15-29.

Mortin, L. I.; and Sedat, J. W. (1982). Structure of *Drosophila* polytene chromosomes. Evidence for a toroidal organization of the bands. *J. Cell Sci.* **57**, 73-113.

Muller, H. J. (1938). The remaking of chromosomes. *Collect. Net.* **13**, 181-195.

Muller, H. J.; and Prokofyeva, A. A. (1935). The individual gene in relation to the chromomere and the chromosome. *Proc. Natl. Acad. Sci. (U.S.A.)* **21**, 16-26.

Nagl, W. (1978). *Endopolyploidy and Polyteny in Differentiation and Evolution.* North-Holland, Amsterdam.

Noll, M. (1974). Subunit structure of chromatin. *Nature* **251**, 249-251.

Nordheim, A.; Pardue, M. L.; Lafer, E. M.; Moller, A.; Stollar, B. D.; and Rich, A. (1981). Antibodies to left-handed Z-DNA bind to interband regions of *Drosophila* polytene chromosomes. *Nature* **294**, 417-422.

Olins, A. L.; and Olins, D. E. (1974). Sheroid chromatin units (V bodies). *Science* **183**, 330-332.

Ostergren, G.; and Praaken, R. (1946). Behaviour on the spindle of the actively mobile chromosome ends of rye. *Hereditas* **32**, 473-494.

Oudet, P.; Gross-Bellard, M.; and Chambon, P. (1975). Electron microscopic and biochemical evidence that chromatin is a repeating unit. *Cell* **4**, 281-300.

Painter, T. S. (1933). A new method for the study of chromosome rearrangements and the plotting of chromosome maps. *Science* **78**, 585-586.

Painter, T. S. (1934). Salivary chromosomes and the attack on the gene. *J. Hered.* **25**, 465-476.

Panitz, R. (1960). Innersekretorische Wirkung auf Strukturmodifikationen der Speicheldrusenchromosomen von *Acricotopus lucidus*. *Naturwissenschaften* **47**, 383.

Panitz, R. (1964). Hormonkontrollierte Genaktivitaten in den Riesenchromosomen von *Acricotopus lucidus*. *Biol. Zb.* **83**, 187-230.

Pardue, M. L.; Kedes, L. H.; Weinberg, E. S.; and Birnstiel, M. L. (1977). Localization of sequences coding for histone messenger RNA in the chromosomes of *Drosophila melanogaster*. *Chromosoma* **63**, 135-153.

Paul, J. (1972). General theory of chromosome structure and gene activation in eukaryotes. *Nature* **238**, 444-446.

Pelling, C. (1959). Chromosomal synthesis of ribonucleic acid as shown by the incorporation of uridine labelled with tritium. *Nature* **184**, 655-656.

Pelling, C. (1964). Ribonukleinsauresynthese der Riesenchromosomen. Autoradiographische Untersuchungen an *Chironomus tentans*. *Chromosoma* **15**, 71-122.

Plagens, U.; Greenleaf, A. L.; and Bautz, E. K. F. (1976). Distribution of RNA polymerase on *Drosophila* polytene chromosomes as studied by indirect immunofluorescence. *Chromosoma* **59**, 157-166.

Poulson, D. F.; and Metz, C. W. (1938). Studies on the structure of nucleolus forming regions and related structures in the giant salivary gland chromosomes of Diptera. *J. Morphol.* **63**, 363-395.

Procunier, J. D.; and Tartof, K. D. (1975). Genetic analysis of the 5S genes in *Drosophila melanogaster*. *Genetics* **81,** 515-523.

Procunier, J. D.; and Dunn, R. J. (1978). Genetic and molecular organization of the 5S locus and mutants in *Drosophila melanogaster*. *Cell* **15,** 1087-1093.

Rambousek, F. (1912). Cytologische Verhaltnisse der Speicheldrusen der *Chironomus*-Larvae. *Sitzungsber. Konig. bohm. Ges. Wiss., math.-naturw. Klasse.* Cited after Beerman (1962).

Renkawitz-Pohl, R.; and Bialojan, S. (1984). A DNA sequence of *Drosophila melanogaster* with a differential telomeric distribution. *Chromosoma* **89,** 206-211.

Risau, W.; Symmons, P.; Saumweber, H.; and Frasch, M. (1983). Nonpackaging and packaging proteins of hnRNA in *Drosophila melanogaster*. *Cell* **33,** 529-541.

Ritossa, F. (1962). A new puffing pattern induced by temperature shock and DNP in *Drosophila*. *Experientia* **18,** 571-573.

Ritossa, F. (1964). Behaviour of RNA and DNA synthesis at the puff level in salivary gland chromosomes. *Exp. Cell Res.* **36,** 515-523.

Rudkin, G. T. (1969). Nonreplicating DNA in *Drosophila*. *Genetics* (suppl.) **61,** 227-238.

Sass, H. (1980). Hierarchy of fibrillar organization levels in the polytene interphase chromosomes of *Chironomus*. *J. Cell Sci.* **45,** 269-293.

Sass, H. (1982). RNA polymerase B in polytene chromosomes: immunofluorescent and autoradiographic analysis during stimulated and repressed RNA synthesis. *Cell* **28,** 269-278.

Sass, H.; and Bautz, E. K. F. (1982a). Immunoelectron microscopic localization of RNA polymerase B on isolated polytene chromosomes of *Chironomus tentants*. *Chromosome* **85,** 633-642.

Sass, H.; and Bautz, E. K. F. (1982b). Interbands of polytene chromosomes: binding sites and start points for RNA polymerase B(II). *Chromosoma* **86,** 77-93.

Saumweber, H.; Symmons, P.; Kabisch, R.; Will, H.; and Bonhoefer, F. (1980). Monclonal antibodies against chromosomal proteins of *Drosophila melanogaster*. *Chromosoma* **80,** 253-275.

Semeshin, V. F.; Zhimulev, I. F.; and Belyaeva, E. S. (1979). Electron microscope autoradiographic study on transcriptional activity of *Drosophila melanogaster* polytene chromosomes. *Chromosoma* **73,** 163-177.

Silver, L. M.; and Elgin, S. C. R. (1976). A method for determination of the *in situ* distribution of chromosomal proteins. *Proc. Natl. Acad. Sci. (U.S.A.)* **73,** 423-427.

Silver, L. M.; and Elgin, S. C. R. (1977). Distribution patterns of three subfractions of *Drosophila* nonhistone chromosomal proteins: possible correlation with gene activity. *Cell* **11,** 971-983.

Silver, L. M.; and Elgin, S. C. R. (1978). Production and characterization of antisera against three individual NH proteins: a case of a generally distributed NH protein. *Chromosoma* **68,** 101-114.

Sirlin, J. L. (1960). Cell sites of RNA and protein synthesis in the salivary gland of *Smittia*. (Chironomidae). *Exp. Cell Res.* **19,** 177-180.

Skaer, R. J. (1977). Interband transcription in *Drosophila*. *J. Cell Sci.* **26,** 251-266.

Sorsa, V. (1974). Organization of replicative units in salivary gland chromosome bands. *Hereditas* **78,** 298-302.

Southern, E. M. (1975). Detection of specific sequences among DNA fragments separated by gel electrophoresis. *J. Mol. Biol.* **98,** 503-518.

Spear, B. B.; and Gall, J. G. (1973). Independent control of ribosomal gene replication in polytene chromosomes of *Drosophila melanogaster*. *Proc. Natl. Acad. Sci.* **70,** 1359-1363.

Spierer, P.; Spierer, A.; Bender, W.; and Hogness, D. S. (1983). Molecular mapping of genetic and chromomeric units in *Drosophila melanogaster*. *J. Mol. Biol.* **168,** 35-50.

Spierer, A.; and Spierer, P. (1984). Similar level of polyteny in bands and interbands of *Drosophila* giant chromosomes. *Nature* **307,** 176-178.

Spradling, A. C.; and Rubin, G. M. (1982). Transposition of cloned P. Elements into *Drosophila* germ line chromosomes. *Science* **218,** 341-347.

Steiner, E. K.; Eissesnberg, J. C.; and Elgin, S. C. R. (1984). A cytological approach to the ordering of events in gene activation using the Sgs-4 locaus of *Drosophila melanogaster*. *J. Cell Biol.* **99,** 233-238.

Swift, H. (1964). The histones of polytene chromosomes. In The *Nucleohistones*. (P. T'so and J. Bonner, Eds.) pp. 169-183, Holden Day, San Francisco.

Tissieres, A.; Mitchell, H. K.; and Tracy, U. M. (1974). Protein synthesis in salivary glands of *Drosophila melanogaster*. Relation to chromosome puffs. *J. Mol. Biol.* **84,** 389-398.

Wang, A. H. J.; Quigley, G. J.; Kolpak, F. J.; van der Marel, G.; van Boom, J. H.; and Rich, A. (1981). Left-handed double helical DNA: variations in the backbone conformation. *Science* **211,** 171-176.

Will, H.; and Bautz, E. K. F. (1980). Immunological identification of a chromocenter-associated protein in polytene chromosomes of *Drosophila*. *Exp. Cell Res.* **125,** 401-410.

Wimber, D. E.; and Steffensen, D. M. (1970). Locatization of 5S RNA genes on *Drosophila* chromosomes by RNA-DNA hybridization. *Science* **70,** 639-645.

Woodruff, R. C.; and Ashburner, M. (1979). The genetics of a small autosomal region of *Drosophila melanogaster* containing the structureal gene for alcohol dehydrogenase. II. Lethal mutations in the region. *Genetics* **92,** 133-149.

Wright, T. R. F.; Beermann, W.; Marsh, J. L.; Bishop, C. P.; Steward, R.; Black, B. C.; Tomsett, A. D.; and Wright, E. Y. (1981). The genetics of dopa decarboxylase in *Drosophila melanogaster*. IV. The genetics and cytology of the 37B10-37D1 region. *Chromosoma* **83,** 45-38.

Young, B. S.; Pession, A.; Traverse, K. L.; French, C.; and Pardue, M. L. (1983). Telomere regions in *Drosophila* share complex DNA sequences with pericentric heterochromatin. *Cell* **34,** 85-94.

Young, M. W.; and Judd, M. H. (1978). Nonessential sequences, genes, and the polytene chromosome bands of *Drosophila melanogaster*. *Genetics* **88,** 723-742.

Zhimulev, I. F.; and Belyaeva, F. A. (1975). 3H uridine labeling patterns in the *Drosophila melanogaster* salivary gland chromosomes X, 2R and 3L. *Chromosoma* **49,** 219-232.

Zhimulev, I. F.; Pokholkova, G. V.; Bgatov, A. V.; Semeshin, V. F.; and Belyaeva, E. S. (1981). Fine cytogenetical analysis of the band 10A1-2 and the adjoining regions in the *Drosophila melanogaster* chromosome. II. Genetical Analysis. *Chromosoma* **82,** 25-40.

Index

ADP-ribosylation, 98-101
5-Azacytidine, 63, 209-212

Centromeres
  lampbrush chromosomes, 158, 164, 166
  metaphase chromosomes, 112-113, 114
  polytene chromosomes, 222
  synaptonemal complexes, 129, 130
Chromatin, transcriptionally active
  DNA methylation, 61-65
  DNase I hypersensitive sites, 65-79
  DNase I sensitivity, 53-58
  DNase II sensitivity, 56
  micrococcal nuclease sensitivity, 55-57
  torsional strain, 60, 72-75
Chromatin domains. See Chromatin loops
Chromatin loops
  DNase I sensitive domains, 54-55
  interphase chromatin, 3, 4, 54-55, 116, 118, 120-121
  lampbrush chromosomes, 154-162, 166-177
  meiotic chromosomes, 137-138
  metaphase chromosomes, 103, 106, 109, 115
  polytene chromosomes, 230-231
Chromatin structure
  chromatosomes, 9-13
  helical ribbon model, 24-30
  hierarchy, 3-4
  historical perspective, 1-3
  ionic strength dependence, 19-21, 23, 25, 105-109
  nucleosomes, 13-17
  polytene chromosomes, 229-231
  preparative methods, 20-21
  solenoid model, 24-30, 93
  superbead model, 21-23
  thick (300 Å) filaments, 19-31, 93, 109, 118
  thin (100 Å) filaments, 4, 11, 17-19
  transcriptionally active chromatin, 52-79, 231
Chromatosome
  composition, 9
  DNA organization, 10-13
  H1 histone organization, 9-13
  orientation in chromatin filaments, 11, 17-19
Chromomeres, 154, 157, 160, 163-165, 174
Chromosome bands
  metaphase chromosomes, 103, 164
  polytene chromosomes, 222, 228-236
Chromosome bouquet, 128, 129
Chromosome pairing. See Synapsis
Chromosome scaffold, 113-116, 118, 138
Core particle
  composition, 5
  DNA organization, 5-6
  histone organization, 6-9

DNA
  PDNA, 143-144
  PsnDNA, 143
  ZDNA, 65, 235-236
  zygDNA, 141-143

DNA methylation
　hypomethylation and transcriptional
　　activation, 61-65, 209-212
　X chromosome inactivation, 209-212
　Z-DNA, 65
DNase I
　sensitivity of
　　active nucleosomes, 57-58
　　transcriptionally active chromatin, 53-55, 75
　hypersensitive sites
　　appearance during transcriptional
　　　activation, 76-79
　　DNA supercoiling, 72-75
　　enhancers, 71
　　fine structure, 69-71
　　functional significance, 68-69
　　locations near active genes, 65-68
　　mechanism of formation, 78-79
　　regulatory proteins, 76
DNA supercoiling
　DNase I hypersensitive sites, 72-75
　DNase I sensitivity of active chromatin, 75
　nucleosomes, 3-13
　thick chromatin fibers, 24-30
　torsional stress and transcription, 60, 72-75
DNA synthesis
　pachytene, 143-144
　premeiotic S phase, 127-128
　zygotene, 141-143
Dosage compensation, 187-188. *See also*
　X-chromosome inactivation

Electric dichroism, 12, 17-18, 26-28
Electron microscopy
　histone-depleted chromosomes, 113-116
　interphase chromatin, 18-22, 25
　lampbrush chromosomes, 165, 171-172
　metaphase chromosomes, 104-112
Enhancers
　DNase I hypersensitive sites, 71
　transcription, 49-50

Histone-depleted chromosomes
　biochemical properties, 113-114
　DNA organization, 114-115, 117
Histone-depleted nuclei, 120-121
Histones
　acetylation, 30, 59
　ADP-ribosylation, 98-99
　H1
　　chromatosomes, 9-13, 18

　　exchange, 20-21
　　linker DNA length, 15
　　polytene chromosomes, 235
　　thick fiber stability, 29-30
　nucleosomes
　　chromatosomes, 9-13
　　core particles, 3, 5, 6-9
　oligomeric complexes, 6-9
　phosphorylation, 94, 97
　transcriptionally active chromatin, 59
　variants, 9
HMG proteins
　DNase I sensitivity of chromatin, 55-59
　lampbrush chromosomes, 173
　polytene chromosomes, 235

Lampbrush chromosomes
　centromeres, 164, 166
　chromomeres, 154, 157, 160, 163-165, 174
　formation, 153, 159-162, 170, 173-174, 180
　functional significance, 177-182
　historical perspective, 152-154
　interchromomeric fibrils, 163
　loops
　　DNA content, 170
　　heterozygosity, 159, 177
　　morphology, 154-162, 166-169
　　proteins, 171-173
　　"read through" hypothesis, 175-179
　　repetitive DNA, 174, 178-179
　　retraction of, 160, 163, 173-174
　　transcription units, 166-169, 174-177, 182
　master-slave hypothesis, 153, 180
　morphology, 154-159
　phylogenetic distribution, 154, 180-181
Linker DNA. *See* Nucleosomes
L protein, 142-143

Meiosis overview, 126-130
Meiotic chromosomes. *See also* Lampbrush
　　chromosomes
　axial cores, 128
　chromatin loops, 137-138
　chromosome bouquet, 128, 129
　DNA metabolism, 141-144
　synapsis, 126-130
　synaptonemal complexes, 128-141
Meiotic endonuclease, 143
Metaphase chromosomes bands, 103, 164
　centromeres, 112-113, 114

comparison to other chromosomes, 103, 119-120, 138, 164
composition
  DNA, 92-93
  histones, 93-94
  nonhistones, 94-101
constancy of structure, 103
folded-fiber model, 111-112, 118
helical-coil model, 118-119
ionic strength dependence, 105-109
isolation, 102-103
nucleosomes, 93-94
radial loops, 106, 109, 115, 116
scaffold, 113-116
scanning electron microscopy, 106-109
telomeres, 113
transmission electron microscopy of sectioned chromosomes, 104-106
whole-mount electron microscopy, 109-112

Nonhistone proteins
  ADP-ribosylation, 98-101
  conservation from interphase to metaphase, 101
  kinetochore proteins, 114, 115
  lamins, 96, 97, 99
  lampbrush chromosomes, 171-173
  metalloproteins, 114, 116, 121
  metaphase chromosomes, 94-101
  phosphorylation, 97-98
  polytene chromosomes, 234-235
  scaffold proteins, 114, 121
  synthesis, 94-96
  two-dimensional gel electrophoresis, 94-101
Nuclear scaffold. *See* Nuclear matrix
Nuclear lamina, 96, 97, 99, 121, 139-141
Nuclear matrix, 96, 120-121, 138-141
Nucleosomes
  chromatosomes
    orientation, 17-19, 24, 27-29
    structure, 9-13
  core particles, 3-9
  historical perspective, 1-3
  HMG proteins, 57-59
  linker DNA, 13-17, 28-29
  metaphase chromosomes, 93-94
  phasing, 15-17, 60-61
  thick chromatin fibers, 19-31
  thin chromatin filaments, 11, 17-19
  transcriptionally active chromatin, 57-60

Phasing. *See* Nucleosomes
Poly (ADP-ribose) polymerase, 98-99. *See also* ADP-ribosylation
Polytene chromosomes
  bands
    chromatin structure, 229-231
    molecular basis of, 228-230
    relationship to genes, 230-231
  chromatin loops, 230
  chromocenter, 22
  differential DNA replication, 229
  general properties, 221-228
  immunofluorescence of proteins, 234-235
  puffs, 231-234
  telomeres, 222
Polytene nuclei, 221-223
Premeiotic S phase
  chromosome pairing, 126-127
  DNA replication, 127-128
Promoters, 48-52, 176-177

Recombination nodules, 131, 133
Psn RNA, 144

Synapsis
  bivalent interlocking, 129
  premeiotic S, 126-127
  sequence during meiotic prophase, 128-130
Synaptic adjustment, 129
Synaptonemal complexes
  assembly and disassembly, 128-129, 133-134, 135, 138
  chemistry, 134-137
  chiasma, 130, 134
  chromatin loops, 138
  chromosome scaffolds, 138
  microspreading methods, 131
  nuclear matrix, 135, 138-141
  polycomplexes, 134
  recombination nodules, 131, 133
  significance to crossing over, 130
  structure, 130-133

Telomeres, 113, 128-129, 222-223
Transcription
  chicken globin genes, 44-46, 62, 77
  hormone responsive genes, 46
  human globin genes, 43-44
  lampbrush chromosomes, 166-171, 174-179, 182
  liver specific proteins, 46-47
  murine globin genes, 41-43, 61, 62, 65, 78

parotid gland specific proteins, 47
polytene chromosomes, 231-235
X-linked genes in *Drosophila,* 188
X-linked genes in mammals
  glucose-6-phosphate dehydrogenase, 189, 207, 209
  hypoxanthine phosphoribosyltransferase, 189, 194, 197-198, 202, 207-209
  phosphoglycerate kinase-1, 190, 199-201, 209
  steroid sulfatase, 191
  Xg blood antigen, 191
Transcriptional regulation
  development, 44-45, 76-78
  DNase I hypersensitive sites, 65-79
  DNA supercoiling, 72-75
  DNA transfection assays, 48-52
  enhancers, 49, 50
  generality, 40-41
  initiation complexes, 50
  promoters, 48, 50, 176-177
Transcription rates
  differentiation specific genes, 43-47
  lampbrush chromosomes, 169
  nascent chain elongation, 39-41
Transcription units
  lampbrush chromosomes, 166-169, 174-177, 182
  polytene chromosomes, 230-231, 233
  single copy genes, 41-47
Topoisomerase I, 59, 74-75, 235

"Unineme" hypothesis, 92, 112

X-chromosome inactivation
  embryogenesis, 192
    ectodermal derivatives, 195-196, 200
    germ cell progenitors, 196
    inner cell mass, 194-195
    trophectoderm, 194-195, 199-201, 205-206
  embryonal carcinoma cells, 194, 206
  evidence for inactivation in females, 189-190
  imprinting, 204-206
  mechanism
    DNA methylation, 209-212
    models, 203
    X chromosome controlling element, 203-206
  nonrandom inactivation, 197
    genetic differences, 197-198
    paternal X chromosome, 199-201, 204-206
    selection versus primary inactivation, 201-202
    X chromosome controlling element, 198, 201, 204-206
  random inactivation, 196-197
  reactivation
    spontaneous, 202-203
    induced, 206-212
  single active X hypothesis, 188-189
  somatic cell hybrids, 207-209
  stability, 202-203, 207-209
X-ray scattering, 2, 10, 21, 29, 93